New Progeria Mutation
Reveals Key Player in
Nuclear Function and Aging

Class

Table of contents

Table of contents

Table of contents

1   Summary

The world's population is ageing rapidly. Currently, people at the age of 60 and over represent 12.3% of the global population. By 2050, this number will rise to almost 22%. Our knowledge of ageing underlying processes is poor, mostly due to the lack of proper models to study it. The straightforward strategy to identify the key genetic players in ageing is the detection of genes leading to accelerated ageing observed in patients with so called progeroid syndromes (PSs), which are very rare genetic disorders characterised by the fatal and severe course of the disease. Clinical features of PSs resemble the physiological processes of ageing, with early manifestation of ageing-associated conditions, such as osteoporosis, atherosclerosis or cancers. Recently, our group examined a 4½-year-old girl diagnosed with a congenital segmental progeria syndrome. Trio-based whole-exome sequencing (WES) led to the identification of a *de novo* variant, located in the *Nuclear pre-lamin A Recognition Factor* (*NARF*) gene. This mutation, c.1100A>G, changed a highly conserved histidine at position 367 to arginine (p.His367Arg), and is predicted to be damaging by several *in silico* prediction programs. *NARF* is an evolutionarily conserved gene that has its homologues in both yeast (*Saccharomyces cerevisiae*, *NAR1*) and nematodes (*Caenorhabditis elegans*, *OXY-4* or *Y54H5A.4*). In mammals, there is an additional homologue of the *NARF* gene named *iron-only hydrogenase-like protein 1* (*IOP1*), also known as *nuclear pre-lamin A recognition factor-like* (*NARFL*). All four proteins are similar to bacterial iron hydrogenases, but they have lost their hydrogenase activity. Functional analysis of the p.His367Arg mutation identified in our patient revealed that it interferes with the nuclear localisation of NARF. I demonstrated that NARF is able to form homodimers, which are probably important for its translocation to the nucleus. I also showed that substitution of the conserved histidine exhibits a dominant negative effect on the wild-type NARF, resulting in complete mislocalisation of both mutated and wild-type NARF to the cytoplasm. This suggests that dimerisation of NARF allows for the generation of unconventional nuclear localisation signal (NLS), and mutation prevents proper conformation and nuclear import. In my studies, I was able to determine the direct interactions of NARF with two proteins: lamin A and CBX5. Both proteins also interact with each other, and have been associated with premature ageing phenotypes. I established that the corresponding mutation causes prolonged DNA repair induced by UV-light lesions in mouse embryonic stem (mES) cells. In addition, I demonstrated impaired proliferation of mES cells carrying the mutation. Both findings could explain the failed attempts at Narf KI mouse line generation in my study. KI mES cell injections into blastocysts gave rise only to low-grade chimeras without germline transmission. I hypothesise that disturbed nuclear transport of the NARF protein and its accumulation in the cytoplasm probably prevent its proper functioning within the nuclear compartment of the cell. Detailed knowledge of the function of the NARF protein in the nucleus is still lacking. The results described within my thesis present another step towards understanding NARF function and the mechanisms underlying ageing and ageing-associated diseases.

2 Introduction

2.1 Ageing processes

2.1.1 Physiological ageing processes

The World Health Organization (WHO) defines ageing as the result of 'accumulation of a wide variety of molecular and cellular damages over time which in turn lead to a gradual decrease in physical and mental capacity, a growing risk of disease, and ultimately, death' (World Health Organization, 2017). According to the United Nations, the worldwide number of people aged 60 years or over was 962 million in 2017. This is more than twice the number of people in this age group in 1980. Moreover, this number is expected to double by 2050, when it is predicted to reach nearly 2.1 billion (United Nations, 2017). Life expectancy has increased and is still increasing dramatically (Flatt and Partridge, 2018; Salomon et al., 2012). This is mostly due to improved environmental conditions, food, water and hygiene, as well as improved medical care leading especially to the reduced impact of infectious diseases owing to antibiotics and vaccinations (Flatt and Partridge, 2018). With the rapid increase in average human life expectancy, there has been a dramatic escalation in age-associated diseases (AADs; (Salomon et al., 2012). Due to the limited understanding of physiological ageing and the molecular mechanisms underlying AADs, modern medicine offers mostly symptomatic treatment. In light of this fact, research on ageing is now the focus of thousands of laboratories that specialise in the fields of genetics, molecular and cellular biology, biochemistry, and behaviour. Rapid advances in the understanding of mechanisms controlling cellular proliferation, differentiation and survival are leading to new insights into the regulation of ageing. Ageing process begins in a single cell and then engages particular organs, eventually leading to the embrace of the entire body along with its interactions with the surrounding environment. Lopez-Otin et al. (2013) categorised the nine commonly occurring hallmarks of ageing (Figure 1). Biological ageing includes cellular and molecular changes, such as genomic instability, telomere attrition, epigenetic alterations, loss of proteostasis, deregulated nutrient sensing, mitochondrial dysfunction, cellular senescence, stem cell exhaustion, and altered intercellular communication (Lopez-Otin et al., 2013). A better understanding of these characteristics separately, as well as interactions between them, can help to develop particular interventions and treatments to pave the way for improved health during ageing in humans.

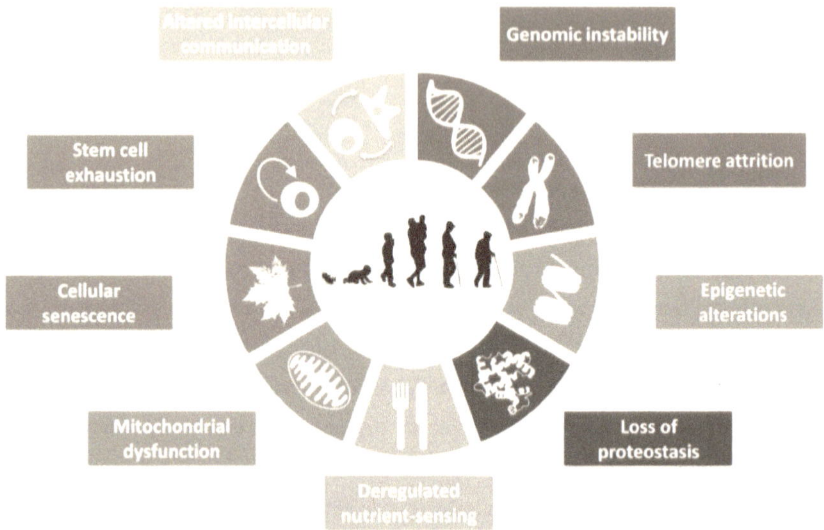

Figure 1: Key molecular hallmarks of the ageing phenotype. The scheme includes the hallmarks of ageing – genomic instability, telomere attrition, epigenetic alterations, loss of proteostasis, deregulated nutrient sensing, mitochondrial dysfunction, cellular senescence, stem cell exhaustion, and altered intercellular communication (adapted from Lopez-Otin et al., 2013).

2.1.2 Accelerated ageing processes

Progeroid syndromes (PSs) are very rare congenital disorders characterised by symptoms of premature or accelerated ageing. PS patients present features that often mimic normal physiological ageing, but they appear at an early age (Carrero et al., 2016; Navarro et al., 2006; Sinha et al., 2014). Typical ageing characteristics observed in PS patients include wrinkled skin, sparse hair, prominent veins, loss of subcutaneous fat, osteoporosis, cardiovascular and neurodegenerative disorders, and cancer (Carrero et al., 2016; Lopez-Otin et al., 2013; Navarro et al., 2006; Sinha et al., 2014). PSs are mostly monogenic disorders, i.e. they are caused by a defect in a single gene. Currently, the Center for Progeroid Syndromes, part of the Center for Rare Diseases Göttingen (ZSEG), offers molecular testing for 85 genes related to 19 known PSs, e.g. Hutchinson-Gilford progeria syndrome (HGPS), Nestor-Guillermo progeria syndrome (NGPS), Wiedemann-Rautenstrauch syndrome (WRS), Werner syndrome (WS), Cockayne syndrome (CS), xeroderma pigmentosum (XP), and Bloom syndrome (BS). Molecular pathways underlying the pathogenesis of PSs include mostly mutations in genes encoding proteins involved in biogenesis and maintenance of the nuclear envelope, i.e. *LMNA* mutations in HGPS or *BANF1* mutations in NGPS, or mutations in genes encoding proteins responsible for various DNA repair mechanisms present in WS, CS, BS or XP (Kubben and Misteli, 2017; Navarro et al., 2006).

2.2 Lamin A alterations cause premature ageing phenotypes

In eukaryotic cells, the nuclear envelope surrounds the nucleus and separates the nuclear and cytosolic cell compartments (Alberts, 2002). The nuclear envelop is composed of an inner and an outer lipid bilayer membrane and nuclear pore complexes, and in metazoan cells, an additional nuclear lamina (Gerace et al., 1984; Gerace and Huber, 2012). The nuclear lamina is connected to the inner nuclear membrane, and provides its mechanical stiffness and stability (Dechat et al., 2010; Dechat et al., 2008; Goldman et al., 2002; Gonzalez et al., 2011). In addition, it takes part in numerous important cellular processes, such as chromatin organisation and remodelling, reorganisation of the nuclear envelope during mitosis, DNA replication, DNA damage repair, cell differentiation, cell migration and transcription control (Dechat et al., 2008; Goldman et al., 2002; Gonzalez et al., 2011). The nuclear lamina consists of V-type intermediate filaments called lamins. According to their sequences, biochemical characteristics and expression profiles, lamins have been qualified as A- and B-type. Mammalian cells express two major forms of B-lamins, B1 and B2, which are encoded by two different genes, *LMNB1* and *LMNB2*. B-type lamins are constitutively expressed, and are substantial for embryonic development (Dechat et al., 2010; Dittmer and Misteli, 2011). In contrast, A-type lamins (lamins A, C, AΔ10 and C2) are products of the alternative splicing of a single gene *LMNA*, and are expressed in differentiated cells (Dechat et al., 2010; Dittmer and Misteli, 2011). All lamins are similar in their domain organisation. They comprise a characteristic central α-helical rod domain, N-terminal head and C-terminal tail domains. The tail domain contains a nuclear localisation signal (NLS), and in lamins B1, B2 and pre-lamin A, an additional CaaX motif in which the 'C' is a cysteine, the two 'a' represant aliphatic amino acids, and the 'X' stands for a random amino acid (Figure 2a). The CaaX motif is subjected to a course of modifications, including farnesylation of the cysteine and proteolytic trimming of aaX by RCE1/ZMPSTE24, followed by carboxymethylation of the farnesylated cysteine. Pre-lamin A is subsequently cleaved again by ZMPSTE24 to remove the additional 15 C-terminal residues (including farnesylated and carboxymethylated cysteine), giving rise to mature lamin A (Figure 2b; Burke and Stewart, 2013; Dittmer and Misteli, 2011).

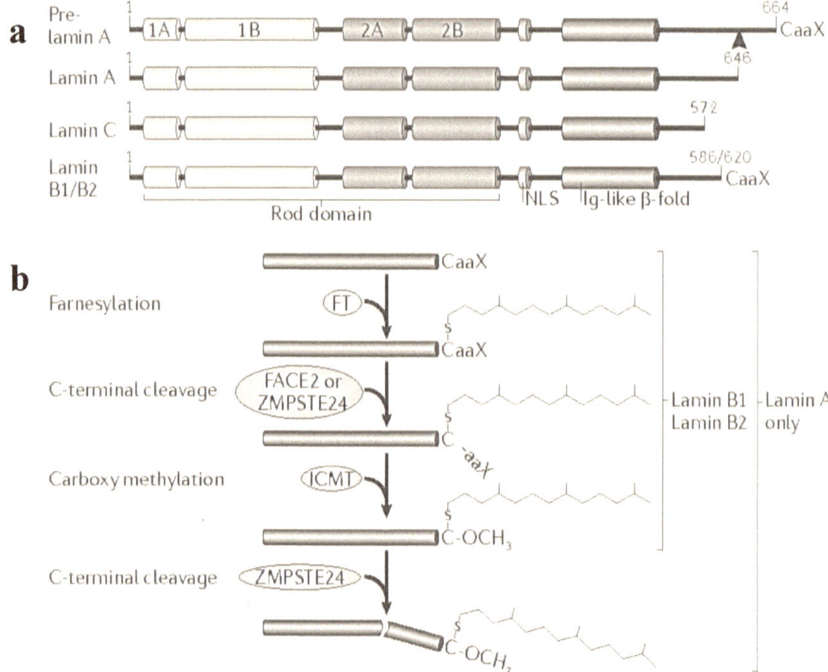

Figure 2: The structure and post-translational modifications of lamins. (a) All lamins are formed of a short N-terminal (head) domain, a central (rod) domain composed of four α-helical domains (1A, 1B, 2A and 2B), and a globular C-terminal (tail) domain that contains lamin-specific motifs: a nuclear localisation signal (NLS), a fold immunoglobulin motif (Ig-like β-fold), and a CaaX motif (C = cysteine; a = aliphatic amino acid; X = any amino acid). **(b)** Lamin B and pre-lamin A undergo post-translational modifications, including farnesylation of the cysteine in the CaaX motif and cleavage by zinc metalloproteinases RCE1 and ZMPSTE24. Farnesylated C is subsequently carboxymethylated by ICMT. Pre-lamin A is further subjected to an additional cleavage of 15 amino acids upstream of the farnesylated/carboxymethylated C, resulting in mature lamin A. FT = farnesyl transferase; ICMT = isoprenylcysteine carboxyl methyltransferase; FACE2 = RCE1 (adapted from Burke and Stewart, 2013).

Mutations in *LMNA* cause a spectrum of human diseases, including Hutchinson-Gilford progeria syndrome (HGPS or progeria, OMIM#176670), the best known and the best characterised progeroid syndrome (De Sandre-Giovannoli et al., 2003; Eriksson et al., 2003). The most common mutation in *LMNA* leading to HGPS is the silent substitution c.1824C>T, p.Gly608Gly, which causes activation of a cryptic exonic splice site that induces altered splicing resulting in a 50-residue truncation of the lamin A protein (Eriksson et al., 2003).

This truncated lamin A (also known as 'progerin') retains the CaaX motif, allowing for farnesylation, while lacking the site for internal proteolytic ZMPSTE24 cleavage. Without the complete processing of pre-lamin A, the resulting progerin accumulates in cells leading to changes in nuclear lamina structure (Figure 3, left panel; (Davies et al., 2009; Eriksson et al., 2003). Mutations in *ZMPSTE24* can also cause accumulation of farnesylated pre-lamin A in cells which results in different progeroid disorders such as restrictive dermopathy (RD; Figure 3, right panel; (Davies et al., 2009). The effects of these mutations underline the importance of the nuclear lamina, and show that alterations in nuclear lamina organisation caused by impaired lamin A processing and maturation may lead to lamina malfunction which is the common cause of severe disorders, such as laminopathies and progeroid syndromes.

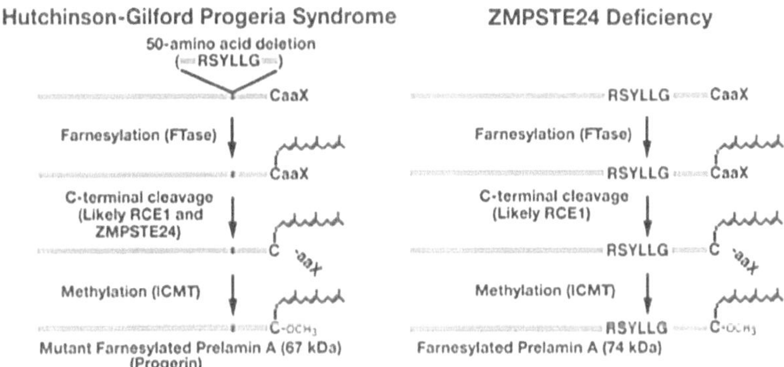

Figure 3: Alterations in lamin A processing. Schemes present alterations in pre-lamin A structure caused either by the mutations in the *LMNA* gene (left) or by mutations in the gene encoding ZMPSTE24 metalloproteinase (right). In patients, both scenarios lead to accumulation of farnesylated pre-lamin A and the progeroid phenotype (adapted from Davies et al. 2010).

2.2.1 Molecular pathogenesis underlying HGPS

Physiological ageing and HGPS share numerous cellular features—such as abnormal nuclear shape, genome instability, telomere attrition, and increased DNA damage—and tissue pathologies, such as diminished bone density and cardiovascular diseases (Burtner and Kennedy, 2010; Vidak and Foisner, 2016). The accumulation of permanently farnesylated progerin within the nuclear membrane is considered to be a toxic attribute underlying the pathogenesis of HGPS (Vidak and Foisner, 2016). Progerin-expressing cells exhibit the altered mechanical properties of the nuclear lamina, culminating in defects in the nuclear architecture, such as thickening, stiffness, and blebbing (Dahl et al., 2006; Goldman et al., 2004). The presence of progerin engenders chromatin disorganisation due to up- and down-regulations of epigenetic modifiers, chromatin regulatory proteins, and proteins of the nucleosome remodelling complexes (Vidak and Foisner, 2016). It has also been shown that progerin expression affects genomic stability via its negative effect on DNA damage repair mechanisms. This includes the impaired recruitment of DSBs, repair proteins involved in homology-directed recombination (HDR), and non-homologues end joining pathways, such as RAD51 and 53BP1, or malfunction of the nucleotide excision repair proteins, such as XPA (Gonzalo and Kreienkamp, 2015; Liu et al., 2005; Vidak and Foisner, 2016). Telomere dysfunction and decreased telomere length are other characteristic features of HGPS cells (Decker et al., 2009). It has been identified that mature lamin A and progerin exhibit different preferences for interacting proteins (Kubben et al., 2010). Disrupted association between progerin and lamin A interacting partner (LAP2α), which ensures proper telomere distribution, engenders telomere mislocalisation, disorganisation, and premature senescence in HGPS cells (Chojnowski et al., 2015). Interestingly, both the reduction of lamin A and the accumulation of farnesylated pre-lamin A result in increased basal and induced levels of reactive oxygen species. These observations are driven by alterations in mitochondrial function, as elevation and accumulation of ROS eventually engender mitochondrial membrane hyperpolarisation and apoptosis. Furthermore, the expression of key ROS-detoxifying enzymes can also be dysregulated (Sieprath et al., 2015).

The progressive increase in knowledge regarding the molecular pathways underlying HGPS pathogenesis has enabled investigation of potential therapeutic approaches for progeria patients. The first successful therapy developed and tested in clinical trials was based on the inhibition of pre-lamin A farnesylation via farnesyltransferase inhibitor (FTI) drug lonafarnib, which was initially developed as an anti-cancer drug. Lonafarnib administration improved the weight gain, the cardiovascular status, and the bone structure of children with HGPS (Gordon et al., 2012; Gordon et al., 2014). Nevertheless, it was demonstrated that the presence of FTI may activate alternative pathways of protein prenylation through geranylgeranylation; therefore, new therapeutic strategies, including combined administration of statins and aminobisphosphonates, have been applied and have shown efficient inhibition of both farnesylation and geranylgeranylation of progerin and pre-lamin A (Varela et al., 2008).

Ongoing studies on potential therapies for patients with HGPS are also considering a reduction of progerin through upregulation of the autophagy with retinoids and rapamycin, an RNA-targeting correction of the splicing defect, and the beneficial effect of resveratrol, which increases the deacetylase activity of SIRT1, a lamin A binding protein (Vidak and Foisner, 2016).

2.3   Identification of novel diseases-causing genes in the era of NGS

2.3.1   Next-generation sequencing (NGS)

In 1977, genome research was revolutionised by the introduction of a new technique for DNA sequencing. Sanger and co-workers described a method of sequencing based on the termination of the PCR reaction by incorporating dideoxynucleotides into the newly synthesised DNA strand (Sanger et al., 1977). It was the beginning of molecular studies that allowed sequencing of DNA, known nowadays as first-generation sequencing (Pareek et al., 2011). Due to the enormous technological development over the years, great progress has been made in DNA sequencing methods. Sanger sequencing requires a considerable amount of time, especially when analysing large genomic regions, and results in high cost. Therefore, there was a concrete need for novel, cost- and time-efficient sequencing methods. The recent introduction of second-generation sequencing or next-generation sequencing (NGS) has led to significant improvement in DNA sequencing. NGS-based methods are faster and cheaper because of much higher throughput by sequencing a large number of DNA strands in parallel (Behjati and Tarpey, 2013). Today, NGS has found numerous applications in many fields of science, including the analysis of genetic diseases and identification of novel disease-causing genes and mutations. The wide availability of different platforms (Garrido-Cardenas et al., 2017; Liu et al., 2012; Pareek et al., 2011) has contributed to the development of many methods for identifying pathogenic mutations, starting from single-gene sequencing, through multi-gene panels designed for particular sets of disorders, to the use of whole-exome sequencing (WES) to search for new, unidentified genetic variants (Rehm, 2013; Saudi Mendeliome, 2015). Despite the fact that NGS is quickly displacing the old sequencing methods, Sanger sequencing is still recommended for confirmation of results obtained by the new generation methods (Mu et al., 2016). Genome sequencing using the massive parallel next-generation sequencing strategies proved to be an effective alternative to locus-specific and gene-panel tests for establishing a new genetic basis of rare diseases (Yang et al., 2013). Protein coding genes determine only 1% of the human genome; however, include approximately 85% of the mutations with disease-related effects (Choi et al., 2009). WES is the next-generation application to identify new variants within all coding regions of known genes. WES coverages more than 95% of the exons, containing disease-causing mutations in Mendelian disorders and many disease-predisposing single nucleotide polymorphisms (SNPs; Rabbani et al. 2013). In my studies, WES technology has been used to identify disease causing variant in *NARF* gene in a patient diagnosed with PS.

2.3.2 Validation of novel disease-causing mutations

NGS has dramatically increased our ability to read information from the human genome, including the DNA sequence, transcriptome and epigenome. Application of these data has improved the identification of new potentially disease-causing genetic variants. To unequivocally recognise new disease-associated genes and to establish the link to disease predisposition and/or progression, NGS data require scientific interpretation (Bonjoch et al., 2019). As a first tool to validate the pathogenicity of variants found (approx. 25.000 in a WES analysis), different bioinformatics approaches have been developed and are currently widely used. Computational methodologies for predicting the effect of mutations include four main categories: sequence conservation, structure analysis, combined (sequence and structure information), and meta-prediction (integrated results from multiple predictors; Tang and Thomas, 2016). Numerous platforms are available which allow prediction if the identified variant may be deleterious, e.g. SIFT (Sorting Intolerant From Tolerant), PANTHER (Protein ANalysis THrough Evolutionary Relationships), PolyPhen-2 (Polymorphism Phenotyping v2), and their combinations such as CAROL (Combined Annotation scoRing tool; SIFT + PolyPhen2; (Bonjoch et al., 2019). Nevertheless, computational bioinformatics analyses alone are not sufficient to claim that an identified variant is a disease-causing mutation. Therefore, all potential variants need to undergo further scientific validation, including detailed variant analysis in different cellular and animal model systems. Flat, two-dimensional (2D) cell culture has dominated scientific research for *in vitro* understanding of the mechanisms of cell behaviour *in vivo*. However, recent studies have shifted towards culture of cells as three-dimensional (3D) structures that ensures more realistic biochemical and biomechanical microenvironments *in vitro* (Duval et al., 2017). Moreover, recent improvements in 3D cell culture techniques allow for creating even more advanced *in vitro* models that are organoids. Both embryonic and adult stem cells can be used to create models of organ development or 'in-a-dish' diseases. Furthermore, organoids based on patient-derived induced pluripotent stem (iPS) cells can be used in numerous applications, such as personalised drug analysis or regenerative medicine, or combined with genome editing methods (see below) in gene therapy

(Clevers, 2016). Despite the fact that the rapidly growing field of *in vitro* research is very helpful in human disease modelling, it is not able to fully replace the benefits of testing animal models *in vivo*. Research on ageing frequently uses small invertebrates such as *Drosophila melanogaster* (Piper and Partridge, 2018) and *Caenorhabditis elegans* (Litke et al., 2018); as models with short lifespans, they are very useful in studying ageing. In addition, there are numerous mouse lines carrying different genetic mutations that act as ageing or ageing-associated disease models (Koks et al., 2016). Interestingly, humans are also investigated in ageing research. Centenarians have been used to study epigenetic signatures of healthy ageing (Puca et al., 2018). It may even be possible to use life in space as a model for ageing, since microgravity causes physiological changes that resemble ageing which are restored after re-entry, allowing investigation of ageing both ways – not only during its development but also during recovery (Biolo et al., 2003). Combining different *in vitro* and *in vivo* models with advancing techniques of genome editing has enabled development of various tools suitable for validation of newly identified genetic variants in ageing processes, progeroid syndromes and other genetic disorders.

2.4   Genome engineering for gene editing approaches

High-throughput techniques of DNA sequencing has allowed the identification of many potential disease-causing genes, and the number of such genes is still growing rapidly. However, the increasing knowledge about disease-causing genetic mutations is not reflected in the development of methods for their treatment. Rare genetic disorders are difficult to diagnose, and despite the existence of many symptomatic treatments, the majority of them are still incurable and often fatal. Currently, gene therapy seems to be the attractive way to treat genetic diseases. Gene therapy is based on the concept of replacing defective DNA with exogenous, correct DNA (Friedmann and Roblin, 1972; Maeder and Gersbach, 2016). A progressive step in gene therapy research was the discovery of the formation of double-strand breaks (DSBs) on DNA and the mechanisms for their repair (Takata et al., 1998; Weaver, 1995). Homology-directed repair (HDR) and non-homologous end joining (NHEJ), which are the two major cellular pathways for repair of DNA DSBs, are widely used today in the gene editing process (Fernandez et al., 2017; Lee et al., 2016b; Maeder and Gersbach, 2016; Takata et al., 1998). Genome editing became faster and easier thanks to the improvement of genetic engineering. Genetically engineered nucleases have become an excellent tool for disrupting harmful genes and introducing changes at the single base-pair level (Fernandez et al., 2017; Lee et al., 2016b; Maeder and Gersbach, 2016).

2.4.1 The history of gene editing methods

Meganucleases are modified naturally occurring homing nucleases (Maeder and Gersbach, 2016). Homing nucleases are small proteins (< 40 kDa) which recognise and cleave specific DNA sequences to form DSBs and induce homologous recombination. They are able to recognise long (14–40 bp) DNA target sites, and are resistant to small changes in these targeted sequences; therefore, they are considered to be the most specific naturally occurring restriction enzymes (Chevalier and Stoddard, 2001; Jurica and Stoddard, 1999; Kowalski and Derbyshire, 2002). The first homing nucleases to be discovered were algal I-*CreI* from *Chlamydomonas reinhardtii* (Heath et al., 1997; Jurica et al., 1998) and yeast PI-*SceI* from *Saccharomyces cerevisiae* (Duan et al., 1997). Both nucleases belong to the large LAGLIDADG family of proteins containing one or two LAGLIDADG motifs (Belfort and Roberts, 1997; Jurica et al., 1998). Currently, genetic engineering is used to manipulate homing nucleases and create engineered meganucleases and chimeric meganucleases that can recognise and process specifically designed target sites in genomes of different organisms (Epinat et al., 2003; Maeder and Gersbach, 2016; Thierry and Dujon, 1992).

Zinc-finger proteins (ZFPs) form the largest family of transcription factors in eukaryotes (Tupler et al., 2001). The first zinc-finger domain was discovered in the *Xenopus laevis* transcription factor IIIA (TFIIIA; (Miller et al., 1985). Because of the ability of zinc-finger domains to recognise and bind to specific DNA sequences, ZFPs make another great tool for gene editing. Combining zinc-finger domains with the cleavage domain of bacterial endonuclease *FokI* obtained from *Flavobacterium okeanokoites* (Li et al., 1992) results in the formation of artificial chimeric zinc-finger nucleases (ZFNs) that are able to create DSBs near to any predesigned DNA sites (Kim et al., 1996; Kim and Chandrasegaran, 1994). However, some of the disadvantages of this method are its long synthesis time and the fact that appropriate ZFN pairs cannot be designed for each genomic locus (Addgene, 2017).

Transcription activator-like effectors (TALEs) have been discovered in the plant pathogenic bacteria *Xanthomonas* spp. (Boch and Bonas, 2010; Deng et al., 2012). In nature, TALEs, which are delivered to plant host cells by bacteria, bind genomic DNA at certain promoter elements leading to activation of expression of genes that are involved in facilitating infections. TALEs recognise specific DNA sequences through their central DNA-binding domain consisting of tandem repeats. Each repeat is made up of 33–35 amino acids (Boch and Bonas, 2010; Deng et al., 2012; Schornack et al., 2008). All tandem repeats have conserved amino acid composition and differ only at position 12 and 13 (repeat-variable di-residues

[RVDs]). RVD composition determines which nucleotide is recognised by a single repeat (Boch et al., 2009; Deng et al., 2012; Moscou and Bogdanove, 2009). Miller and colleagues combined synthetic engineered versions of TALEs with the cleavage domain of *Fok*I endonucleases (similarly to ZFNs) resulting in new engineered nucleases – transcription activator-like effector based nucleases (TALENs; (Miller et al., 2011). An additional advantage of this method is the shorter synthesis time of TALENs compared to that of ZFNs (Addgene, 2017; Lee et al., 2016b). However, increasing specificity by extending repetitive segments also increases the size of the nuclease, making it difficult to introduce TALENs into cells (Lee et al., 2016b; Maeder and Gersbach, 2016; Rinaldi et al., 2017).

2.4.2 CRISPR/Cas technology as a novel gene editing tool

Today, the clustered regularly interspaced short palindromic repeats (CRISPR)/CRISPR-associated (Cas) system is the most commonly used technique for gene editing. CRISPR were discovered in *Escherichia coli* in 1987 (Ishino et al., 1987), and have been studied extensively since then. CRISPR and CRISPR-associated genes (Cas) form an efficient bacterial resistance system which protects them against bacteriophage invasion (Barrangou et al., 2007; Garneau et al., 2010; Terns and Terns, 2011). Twenty-five years after the discovery of bacterial CRISPR/Cas, the first report on the benefits of this system for editing eukaryotic genomes appeared. Based on this system, researchers created a simple complex of CRISPR/Cas9 that allows recognition of very specific DNA sequences activating DNA cleavage. The DSBs produced are then repaired by NHEJ or HDR (Addgene, 2017; Jinek et al., 2012). The native CRISPR/Cas systems use two RNAs: CRISPR RNA (crRNA) that guides nucleases to specific places in the genome and trans-activating crRNA (tracrRNA) that serves as a scaffold for Cas and crRNA and also participates in the maturation of crRNA from its precursor form pre-crRNA (Addgene, 2017; Bhaya et al., 2011; Brouns et al., 2008; Deltcheva et al., 2011; Wiedenheft et al., 2012). In engineered CRISPR/Cas9 systems, RNAs are simplified and condensed into single shorter guide RNA (gRNA; (Addgene, 2017; Fu et al., 2014; Jinek et al., 2012). The gRNA is composed of a 20-nucleotide sequence complementary to the genomic target and the scaffolding sequence necessary for binding to Cas9. Shorter gRNAs have reduced levels of off-target events induced by CRISPR/Cas9 (Addgene, 2017; Fu et al., 2014). Cas9 proteins are nucleases that specifically cleave DNA, determined by both complementarity between gRNAs and targeted DNA sites and the presence of a protospacer-adjacent motif (PAM) localised downstream of the target sequence (Jinek et al., 2012; Sapranauskas et al., 2011). PAM is a short (3–8 bp) sequence functioning as a signal for Cas proteins to bind and cleave double-stranded DNA, and it is distinct for particular nucleases (i.e. the most commonly used Cas9 originating from *Streptococcus pyogenes* recognises the 5'-NGG-3' PAM sequence; (Addgene, 2017; Jinek et al., 2012; Mojica et al., 2009). Cas9, brought by the gRNA to the appropriate DNA sequence, recognises PAM, attaches to the DNA and cleaves 3–4 nucleotides upstream of the PAM sequence, resulting in DNA DSBs (Addgene, 2017; Jinek et al., 2012; Sapranauskas et al., 2011). CRISPR/Cas9 can be used to modify any desired genomic target as long as the sequence is unique and upstream of the PAM sequence. Therefore, a need for more PAM sequences arose, and to solve this problem, different variants of modified Cas9 and its homologues are currently used, which, in combination with unique gRNAs, give the ability to edit any sequence in the genome (Addgene, 2017; Kleinstiver et al., 2015). Compared to ZFNs and TALENs, the CRISPR/Cas9 system is easier and faster to synthesise and use, and has become the most popular gene-editing tool (Lee et al., 2016b; Maeder and Gersbach, 2016). Less than four years after the first report on the use of CRISPR/Cas 9 for gene editing, 2600 CRISPR-related publications have appeared in the PubMed database (Addgene, 2017). In my studies, the CRISPR/Cas9 system has been applied to generate knock-in (KI) mouse model of progeroid syndrome.

2.5 *Nuclear pre-lamin A recognition factor (NARF)*

Initially, NARF was identified as a binding partner of farnesylated pre-lamin A (Barton and Worman, 1999). NARF is a 456-amino acid protein, with an expected molecular mass of about 52 kDa, encompassing two protein domains characteristic for iron-only hydrogenases: iron-hydrogenase large and small subunits (Figure 4). NARF does not display hydrogenase activity, and is supposed to develop new functions during evolution (Hackstein, 2005). Barton and Worman (1999) demonstrated that NARF is a nuclear protein which interacts exclusively with the C-terminal tail of prenylated pre-lamin A. NARF does not bind to mature lamin A or prenylated lamin B1. Prenylation (farnesylation) of cysteine in the CaaX motif, but no carboxymethylation of pre-lamin A, seems to be required for NARF binding, and enhances this interaction (Barton and Worman, 1999).

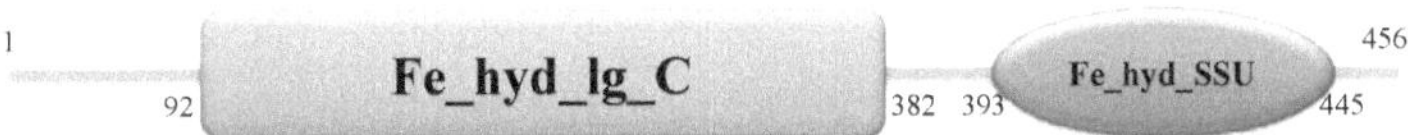

Figure 4: Predicted structure of nuclear pre-lamin A recognition factor (NARF). The scheme presents expected domain organisation in the human NARF protein (prediction with ebi.ac.uk – HMMER tool). The prediction tool distinguishes two domains characteristic of iron hydrogenases: large (green rectangle) and small (blue ellipse) subunits of iron hydrogenase. They extend from amino acids 92–382 and 393–445, respectively (Fe_hyd_lg_C = iron-only hydrogenase large subunit, C-terminal domain; Fe_hyd_SSU = iron hydrogenase small subunit).

The *NARF* gene is quite conserved, and it has its homologues in other eukaryotes, e.g. in yeast *S. cerevisiae* (*NAR1*) and in the nematode *C. elegans* (*OXY-4* or *Y54H5A.4*). Both Nar1 and oxy-4 proteins have been described as hydrogenase-like proteins (Balk et al., 2004; Fujii et al., 2009). In anaerobic prokaryotic cells, hydrogenases are required for the production and metabolism of molecular hydrogen (Peters, 1999). In contrast, yeast Nar1 does not function as a hydrogenase, but it takes part in biogenesis of cytosolic and nuclear iron–sulphur (Fe/S) proteins (Balk et al., 2004). Moreover, Fujii et al. (2009) suggested a potential role for Nar1 and oxy-4 in the regulation of oxidative stress. *nar1* and *oxy-4* mutants showed increased sensitivity to higher concentrations of oxygen in the environment, which resulted in decreased longevity and growth retardation. Interestingly, mutants are sensitive to oxidative damage even under normal oxygen culture conditions (Fujii et al., 2009). In mammals, there is an additional homologue of NARF, namely iron-only hydrogenase-like protein 1 (IOP1) or nuclear pre-lamin A recognition factor-like (NARFL) protein. NARFL has been described as an element involved in the biogenesis of cytosolic iron–sulphur proteins (Huang et al., 2007; Song and Lee, 2008, 2011; Song et al., 2009). The *Narfl* knock-out animal model is characterised by early embryonic lethality, while gene inactivation in adults resulted in a significantly decreased level of cytosolic Fe/S protein leading to premature death (Song and Lee, 2011). In addition, NARFL has been described as a protein involved in the regulation of hypoxia-inducible factor-1α (HIF-1α) activity. Knockdown of this protein in mammalian cells leads to upregulation of HIF-1α under both normal and decreased oxygen conditions, which in turn leads to HIF-1α target gene expression (Huang et al., 2007). Taking together, NARFL participates in cellular respiration, the production of free radicals and oxidative stress, thereby regulating the ageing processes of multicellular organisms (Harman, 2003).

3   Aim of the study

The main aim of this study was the characterisation of a mutation in *Nuclear pre-lamin A Recognition Factor* (*NARF*), associated with a novel progeroid syndrome. The mutation in *NARF* was recently identified by our group in a patient diagnosed with a segmental congenital progeroid syndrome. Due to the lack of knowledge about the functions of NARF, I began my research with basic characterisation of the function of the protein in the cell. I intended to establish subcellular localisation of both wild-type and mutated NARF proteins. In addition, I wanted to identify partners interacting with NARF and confirm direct interactions, using different functional *in vitro* approaches. I aimed to determine the functional similarity of NARF to its homologues by complementation assays performed in a yeast model. Since NARF homologues play a role as key regulators of the oxidative stress response, I attempted to verify their function in reactive oxygen species (ROS) generation. Genomic instability and impaired DNA damage responses are essential hallmarks of ageing; therefore, I aimed to examine the impact of mutation on DNA repair mechanisms by induction of DNA lesions in cells. Finally, I tried to generate an animal model for the identified mutation. To accomplish this, I planned to use the CRISPR/Cas9 system to introduce the mutation corresponding to that of the patient into the mouse genome and generate a knock-in mouse line that could serve as a new model for progeroid syndromes studies. Taking all thesis aims together, this study intended to bring new insights into the pathomechanism underlying the progeroid phenotype presented in affected patient, as well as a general insights into the functions of the NARF protein.

4 Materials and methods

4.1 Materials

4.1.1 Chemicals

Table 1: List of chemicals used during research.

Chemical	Producer
Agar	Carl Roth GmbH, Karlsruhe, Germany
Agarose	Peqlab Biotechnologie GmbH, Erlangen, Germany
Ampicillin	Carl Roth GmbH, Karlsruhe, Germany
Ampuwa (distilled H_2O)	Frensenius Kabi, Bad Homburg, Germany
ß-mercaptoethanol	Gibco™/Thermo Fisher Scientific, Carlsbad, USA
Bacto™ Peptone	BD Bioscience, New Jersey, USA
Bacto™ Yeast Extract	BD Bioscience, New Jersey, USA
CellTiter 96® AQ$_{ueous}$ One Solution Reagent	Promega, Madison, USA
Chloroform	Applichem, Darmstedt, Germany
Clarity™ Western ECL Substrate	Bio-Rad Laboratories Inc., Hercules, USA
Dimethyl sulfoxide (DMSO)	Applichem, Darmstedt, Germany
DNA ladder (1kb/100bp)	Thermo Fisher Scientific, Carlsbad, USA
DNA Loading Dye (6x)	Thermo Fisher Scientific, Carlsbad, USA
dNTPs Mix (10mM)	Thermo Fisher Scientific, Carlsbad, USA
Dulbecco's Modified Eagle Medium (DMEM)	Gibco™/Thermo Fisher Scientific, Carlsbad, USA
Ethanol	J.T. Baker/Thermo Fisher Scientific, Carlsbad, USA
Etoposide	Sigma-Aldrich/Merck, Darmstedt, Germany
Formaldehyde	Merck, Darmstadt, Germany
Fetal bovine serum (FBS Superior)	Merck, Darmstadt, Germany
GelRed® Nucleic Acid Gel Stain	Biotium, Aachen, Germany
Halt™ Protease Inhibitor Cocktail (100x)	Thermo Fisher Scientific, Carlsbad, USA
HEPES	Sigma-Aldrich/Merck, Darmstedt, Germany
HisPur™ Cobalt Resin	Thermo Fisher Scientific, Carlsbad, USA
Hydrogen chloride (HCl)	J.T. Baker/Thermo Fisher Scientific, Carlsbad, USA
Imidazole	Carl Roth GmbH, Karlsruhe, Germany
Isopropanol	J.T. Baker/Thermo Fisher Scientific, Carlsbad, USA
Isopropyl ß-D-1-thiogalactopyranoside (IPTG)	Carl Roth GmbH, Karlsruhe, Germany
L-Glutamine	Gibco™/Thermo Fisher Scientific, Carlsbad, USA
Laemmli Sample Buffer (4x)	Bio-Rad Laboratories Inc., Hercules, USA
Leukemia Inhibitor Factor (LIF)	Millipore/Merck, Darmstedt, Germany

(ESGRO®mLIF)	
Lipofectamine®2000 Reagent	Invitrogen™/Thermo Fisher Scientific, Carlsbad, USA
MES monohydrate	Sigma-Aldrich/Merck, Darmstedt, Germany
Methanol	J.T. Baker/Thermo Fisher Scientific, Carlsbad, USA
Milk powder	Carl Roth GmbH, Karlsruhe, Germany
Mitomycin C	Sigma-Aldrich/Merck, Darmstedt, Germany
Sodium chloride (NaCl)	Sigma-Aldrich/Merck, Darmstedt, Germany
Sodium hydroxide (NaOH)	Merck, Darmstedt, Germany
Sodium phosphate (Na_2HPO_4)	Sigma-Aldrich/Merck, Darmstedt, Germany
Non-essential amino acids (NEAA)	Gibco™/Thermo Fisher Scientific, Carlsbad, USA
OPTI-MEM® I Reduced Serum Medium	Invitrogen™/Thermo Fisher Scientific, Carlsbad, USA
Pansera ES (special designed bovine serum for embryonal stem cells)	PAN-Biotech GmbH, Aidenbach, Germany
Paraformaldehyde	Merck, Darmstadt, Germany
Penicillin/Streptomycin	Gibco™/Thermo Fisher Scientific, Carlsbad, USA
Peptone	Carl Roth GmbH, Karlsruhe, Germany
Phosphate buffer Saline (PBS)	PAN-Biotech GmbH, Aidenbach, Germany Gibco™/Thermo Fisher Scientific, Carlsbad, USA
PhosSTOP EASYpack Phosphatase Inhibitor Cocktail Tablets	Hoffmann-La Roche, Basel, Switzerland
Pierce® RIPA Buffer	Thermo Fisher Scientific, Carlsbad, USA
Precision Plus Protein™ All Blue Standards	Bio-Rad Laboratories Inc., Hercules, USA
ProLong™ Diamond Antifade Mountant with DAPI	Thermo Fisher Scientific, Carlsbad, USA
Reducing Agent (10x)	Invitrogen™/Thermo Fisher Scientific, Carlsbad, USA
Restore™ PLUS Western Blot Stripping Buffer	Thermo Fisher Scientific, Carlsbad, USA
SeeBlue® Plus2 Pre Stained Standard	Invitrogen™/Thermo Fisher Scientific, Carlsbad, USA
TRIzol® Reagent	Invitrogen™/Thermo Fisher Scientific, Carlsbad, USA
Trihydroxymethylaminomethane (Tris)	AppliChem GmbH, Darmstadt, Germany
Tris/Glycin/SDS running buffer (10x)	Bio-Rad Laboratories Inc., Hercules, USA
Triton-X-100	Carl Roth GmbH, Karlsruhe, Germany
Tween®20	Promega, Mannheim, Germany
UltraPure™ TBE Buffer (10x)	Invitrogen™/Thermo Fisher Scientific, Carlsbad, USA
Yeast extract	Carl Roth, Karlsruhe, Germany

4.1.2 Enzymes

Table 2: List of enzymes used during research.

Enzyme	Producer
0,05% Trypsin-EDTA	Gibco™/Thermo Fisher Scientific, Carlsbad, USA
Exonuclease I	New England Biolabs, Ipswich, USA
Platinum® Taq DNA Polymerase	Invitrogen™/Thermo Fisher Scientific, Carlsbad, USA
Proteinase K	Carl Roth, Karlsruhe, Germany
Restriction enzymes (Fast Digest)	Invitrogen™/Thermo Fisher Scientific, Carlsbad, USA
Shrimp Alkaline Phosphatase (SAP)	Promega, Madison, USA
T4 DNA Ligase	Invitrogen™/Thermo Fisher Scientific, Carlsbad, USA

4.1.3 Vectors

Table 3: List of vectors used during research.

Vector	Producer
pCMV-Myc-N	Clontech Laboratories Inc., USA
hEF1α-GFP	Kind gift of Dr. Jessica Nolte, Institute of Human Genetics, UMG, Göttingen, Germany
pCR™4Blunt-TOPO®	Thermo Fisher Scientific, Carlsbad, USA
pCR™II-Blunt-TOPO®	Thermo Fisher Scientific, Carlsbad, USA
pCR™2.1-TOPO®	Thermo Fisher Scientific, Carlsbad, USA
pJET1.2	Thermo Fisher Scientific, Carlsbad, USA
pET 28a (+)	Merck, Darmstadt, Germany
p415-BFP2	Kind gift of Prof. Blanche Schwappach, Department of Molecular Biology, UMG, Göttingen, Germany
p416	Kind gift of Prof. Blanche Schwappach, Department of Molecular Biology, UMG, Göttingen, Germany
p415-ZZ-tag-TEV	Kind gift of Prof. Blanche Schwappach, Department of Molecular Biology, UMG, Göttingen, Germany
pCSDest C-VC, C-VN, N-VC, N-VN	Kind gift of Dr. Roland Dosch, Department of Developmental Biochemistry, UMG, Göttingen, Germany

4.1.4 Primers

All synthetic oligonucleotides were either purchased from Eurofins Genomics (Ebersberg, Germany).

Table 4: List of primers used during research.

Name	Sequence 5'→ 3'	Application
CBX5_EcoRI_F_Myc	GAATTCGGATGGGAAAGAAAACCAAGCGGA	Cloning into pCMV-Myc-N vector
CBX5_XhoI_R_Myc	CTCGAGTTAGCTCTTTGCTGTTTCTTTCTC	
CBX5_EcoRI_F_His	GAATTCATGGGAAAGAAAACCAAGCGGACA	Cloning into pET 28a (+) vector
CBX5_XhoI_R_His	CTCGAGGCTCTTTGCTGTTTCTTTCTCTTT	
hNARF_F_BglII	AGATCTCCATGAAGTGTGAGCACTGCACGCGCAAGGAATGTAGTAAG	Human NARF mutagenesis – patient mutation and cloning into pCMV-Myc-N vector
hNARF_Mut_R	ACAGGCGAGGACCTCCACAAAGCGGAATGGGAACTTGCCCTTCTTAAG	
hNARF_Mut_F	CTTAAGAAGGGCAAGTTCCCATTCCGCTTTGTGGAGGTCCTCGCCTGT	
hNARF_F_XhoI	CTCGAGTCACCACTTGATGTCCAGGCTGTGTGTGCCACGCTCCTG	
mNarf_F_BglII	AGATCTCCATGAAGTGTGAGCACTGCACACGAAAGGAATGTAGTAAA	Mouse Narf mutagenesis – patient corresponding mutation and cloning into pCMV-Myc-N vector
mNarf_Mut_R	ACACGCGAGCACCTCCACAAAGCGGTATGGGAGTTTGCCCTTCTTGAG	
mNarf_Mut_F	CTCAAGAAGGGCAAACTCCCATACCGCTTTGTGGAGGTGCTCGCGTGT	
mNarf_R_XhoI	CTCGAGTCACCACTTGATATCCAGGCCGTCTGTGCAGGGCTCCAA	
hEF1a_NARF_KpnI	GGTACCATGAAGTGTGAGCACTGCACGCGC	Cloning into hEF1α-GFP vector
hEF1a_NARF_BamHI	GGATCCGCCCACTTGATGTCCAGGCTGTGTGTGCC	
NARF_mut1_F	CCTTTGTGGAGGTCCTCGCCTGTG	mutagenesis
NARF_mut1_R	GGGAATGGGAACTTGCCCTTCTTA	
NARF_mut1_F1	GTTCCCATTCCCCTTTGTGGAGGT	mutagenesis
NARF_mut1_R1	CCTCCACAAAGGGGAATGGGAACT	
NARF_mut2_F	TCTTTGTGGAGGTCCTCGCCTGTG	mutagenesis
NARF_mut2_R	AGGAATGGGAACTTGCCCTTCTTA	
NARF_mut2_F1	GTTCCCATTCCTCTTTGTGGAGGT	mutagenesis
NARF_mut2_R1	CCTCCACAAAGAGGAATGGGAACT	
NARF_mut3_F	GACTTTGTGGAGGTCCTCGCCTGT	mutagenesis
NARF_mut3_R	CGAATGGGAACTTGCCCTTCTTAA	
NARF_mut4_F	GTTTGTGGAGGTCCTCGCCTGTGC	mutagenesis
NARF_mut4_R	CTGGAATGGGAACTTGCCCTTCTT	
NARF_mut5_F	TACTTTGTGGAGGTCCTCGCCTGT	mutagenesis
NARF_mut5_R	AGAATGGGAACTTGCCCTTCTTAA	
NARF_mut6_F	TTTTGTGGAGGTCCTCGCCTGTGC	mutagenesis
NARF_mut6_R	ATGGAATGGGAACTTGCCCTTCTT	
NARF_mut6_F1	AAGTTCCCATTCCATTTTGTGGAG	mutagenesis
NARF_mut6_R1	GACCTCCACAAAATGGAATGGGAA	

mNarf_int9-10_F	GGTCTGTGGCATACATGCAG	genotyping
mNarf_int10-11_R	CACTGTTCCCTTCCCTGTGT	
mNarf_ex10_seq F	GCTCTACCTGTGGCTGTTCC	Sequencing
mNarf_ex10_seq R	AACCTTCATGGCTGAGGATG	Sequencing
Narf_HRD_F	GAGAAGAACGGGGAGGTCCTA	Amplification of HRD sequences
Narf_HRD_R	TGGCGTGGGGGAACATCCTGC	
Narf_HRD_Mut_F	TCGCTTTGTGGAGGTGCTCGCGTGTCC	
Narf_HRD_Mut_R	CGATATGGGAGTTTGCCCTTCTTGAGC	
Narf_HRD_WT_F	TCACTTTGTGGAGGTGCTCGCGTGTCC	
Narf_HRD_WT_R	ATATGGGAGTTTGCCCTTCTTGAGCTT	
NARF_BamHI_F	GGATCCATGAAGTGTGAGCACTGCACGCGC	Cloning into pET 28a (+) vector
NARF_XhoI_R-stop	CTCGAGCCACTTGATGTCCAGGCTGTGTGT	
NARF_yopt_NcoI_F2	CCATGGGGATGAAATGCGAACACTGCACAAGA	Cloning into yeast expressing vectors
NARF_yopt_XhoI_R	TCTAGAATGAAATGCGAACACTGCACAAGA	
NARF_yopt_Fseq	CAATCATTGCCATACTTCGC	Sequencing
NARF_yopt_Rseq	CAAGATCATGTTTTGGATGT	Sequencing
NARFL_yopt_NcoI_F2	CCATGGGGATGGCTTCCCCTTTTTCCGGTGCT	Cloning into yeast expressing vectors
NARFL_yopt_XhoI_R	CTCGAGTTACCATCTGATACCCAAACCTGT	
NARFL_yopt_Fseq	AATAAGATGGCTGCACCATC	Sequencing
NARFL_yopt_Rseq	TCTCTTCAATCTTTGAACCA	Sequencing
yNar1_NcoI_F2	CCATGGGGATGAGTGCTCTACTGTCCGAGTCT	Cloning into yeast expressing vectors
yNar1R_XhoI	CTCGAGTTACCAGGTGCTCCCAACAGAGAC	
yNar1_Fseq	GGAGGCGGCCGATTTGTGTT	Sequencing
yNar1_Rseq	CGGTGATGTTCCTCTTGCGC	Sequencing
Nar1_DamP_F1	CTCACTTGATAACCTTATTT	
Nar1_DamP_R1	CGGGCTTTTAGCACCGTTGG	
Nar1_DamP_F2	TGGTGGCCTACTCAATGGCG	
Nar1_DamP_R2	TCAGCCAGTTTAGTCTGACC	
VNC_hLMNA_	GGGGACAAGTTTGTACAAAAAAGCAGGCTTAAT	Gateway

F	GGAGACCCCGTCCCAGCGGCGCGCCACC	cloning
VN_hLMNA_R	GGGGACCACTTTGTACAAGAAAGCTGGGTTTTA CATGATGCTGCAGTTCTGGGGGCTCTG	
VC_hLMNA_R	GGGGACCACTTTGTACAAGAAAGCTGGGTTCAT GATGCTGCAGTTCTGGGGGCTCTGGGT	
VNC_hNARF_F	GGGGACAAGTTTGTACAAAAAAGCAGGCTTAAT GAAGTGTGAGCACTGCACGCGCAAGGAA	Gateway cloning
VN_hNARF_R	GGGGACCACTTTGTACAAGAAAGCTGGGTTTCA CCACTTGATGTCCAGGCTGTGTGTGCC	
VC_hNARF_R	GGGGACCACTTTGTACAAGAAAGCTGGGTTCCA CTTGATGTCCAGGCTGTGTGTGCCACG	
VNC_hCBX5_F	GGGGACAAGTTTGTACAAAAAAGCAGGCTTAAT GGGAAAGAAAACCAAGCGGACAGCTGAC	Gateway cloning
VN_hCBX5_R	GGGGACCACTTTGTACAAGAAAGCTGGGTTTTA GCTCTTTGCTGTTTCTTTCTCTTTGTT	
VC_hCBX5_R	GGGGACCACTTTGTACAAGAAAGCTGGGTTGCT CTTTGCTGTTTCTTTCTCTTTGTTTTC	
qhNARF_F	AAGGGCAAGTTCCCATTCCA	qRT-PCR
qhNARF_R	TTATCCGCATGTCCGTCTGG	
qmNarf_F	TCACGTTTTCAGACACGCAG	qRT-PCR
qmNarf_R	ACCTCCCCGTTCTTCTCAAG	
qmGapdh_F	TCGTCCCGTAGACAAAATGG	qRT-PCR
qmGapdh_R	TTGAGGTCAATGAAGGGGTC	

4.1.5 Antibodies

Table 5: List of antibodies used during research.

Type	Name	Application (dilution)	Producer
Primary antibody	α-tubulin	WB (1:1000)	Sigma-Aldrich/Merck, Darmstedt,Germany
	β-actin	WB (1:1000)	Sigma-Aldrich/Merck, Darmstedt,Germany
	Phospho-Histone H2A.X (Ser139), clone JBW301	WB (1:1000)	Millipore/Merck, Darmstedt, Germany
	Histone H2A.X	WB (1:1000)	Millipore/Merck, Darmstedt, Germany
	Penta-His tag	WB (1:1000)	Thermo Fisher Scientific, Carlsbad, USA
	HP1-α	WB (1:1000) ICC (1:100)	Abcam, Cambridge, UK
	Lamin A+C [EPR4068]	WB (1:1000) ICC (1:100)	Abcam, Cambridge, UK
	Mouse IgG – isotype control	CoIP (10 μg)	Abcam, Cambridge, UK
	Myc Tag, clone 4A6	WB (1:1000)	Millipore/Merck, Darmstedt,

		ICC (1:100)	Germany
Secondary antibody	Alexa Fluor® 488 mouse IgG	ICC (1:300)	Life Technologies/Thermo Fisher Scientific, Carlsbad, USA
	Alexa Fluor® 555 mouse IgG	ICC (1:300)	Life Technologies/Thermo Fisher Scientific, Carlsbad, USA
	Alexa Fluor® 488 rabbit IgG	ICC (1:300)	Life Technologies/Thermo Fisher Scientific, Carlsbad, USA
	Alexa Fluor® 546 rabbit IgG	ICC (1:300)	Life Technologies/Thermo Fisher Scientific, Carlsbad, USA
	Goat IgG-Cy3	ICC (1:300)	Sigma-Aldrich/Merck, Darmstedt,Germany
	Goat IgG-HRP	WB (1:10000)	Sigma-Aldrich/Merck, Darmstedt,Germany
	Mouse IgG-HRP	WB (1:10000)	Sigma-Aldrich/Merck, Darmstedt,Germany
	Rabbit IgG-HRP	WB (1:10000)	Sigma-Aldrich/Merck, Darmstedt,Germany

4.1.6 Kits

Table 6: List of kits used during research.

Kit	Producer
Big Dye Terminator V1.1 and V3.1 Cycle Sequencing Kit	Life Technologies/Thermo Fisher Scientific, Carlsbad, USA
CellROX® Deep Red Flow Cytometry Assay Kit	Life Technologies/Thermo Fisher Scientific, Carlsbad, USA
CloneJET™ PCR Cloning Kit	Thermo Fisher Scientific, Carlsbad, USA
DNase I, Amplification Grade	Thermo Fisher Scientific, Carlsbad, USA
First Strand cDNA Synthesis Kit	Thermo Fisher Scientific, Carlsbad, USA
Gateway® Technology	Life Technologies/Thermo Fisher Scientific, Carlsbad, USA
NE-PER™ Nuclear and Cytoplasmic Extraction Kit	Thermo Fisher Scientific, Carlsbad, USA
NucleoBond® Xtra Maxi Kit	Macherey-Nagel GmbH & Co. KG, Düren, Germany
Nucleo Spin® Plasmid	Macherey-Nagel GmbH & Co. KG, Düren, Germany
Pierce® BCA Protein Assay Kit	Thermo Fisher Scientific, Carlsbad, USA
QIAquick Gel Extraction Kit	Qiagen GmbH, Hilden, Germany
QuantiNova SYBR® Green RT-PCR Kit	Qiagen GmbH, Hilden, Germany
RTA Transfer Kit (Mini, PVDF)	Bio-Rad Laboratories Inc., Hercules, USA
TOPO® Cloning Kits	Thermo Fisher Scientific, Carlsbad, USA

4.1.7 Laboratory materials

Table 7: List of expendable laboratory equipment used during research.

Material	Producer
1,5 ml tubes	Sarstedt, Nümbrecht, Germany
2 ml tubes	Sarstedt, Nümbrecht, Germany
384-well PCR plates	Thermo Fisher Scientific, Carlsbad, USA
Centrisart® I (20,000 MWCO CTA)	Sartorius AG, Göttingen, Germany
Cover slips	Thermo Fisher Scientific, Carlsbad, USA
CryoTube™ Vials	Thermo Fisher Scientific, Carlsbad, USA
Disposable filters	Sartorius AG, Göttingen, Germany
Falcon tubes (15 ml and 50 ml)	Greiner Bio-One, Kremsmünster, Austria
Glass Pasteur Pipettes	Brand GmbH, Wertheim. Germany
Microscope slides	Thermo Fisher Scientific, Carlsbad, USA
Mini-PROTEAN® Precast Gels	Bio-Rad Laboratories, Inc., Hercules, USA
Mitsubishi Thermal Paper Standard KP61B	Biometra GmbH, Göttingen, Germany
PCR tubes	Nippon Genetics Europe GmbH, Düren, Germany
Pipette tips (WB gel loading)	Biozym Scientific GmbH, Hessisch Oldendorf, Germany
Pipette tips (white long)	STARLAB International GmbH, Hamburg, Germany
Pipette tips (qRT-PCR)	STARLAB International GmbH, Hamburg, Germany
Pipette tips (yellow, blue, white short)	Sarstedt, Nümbrecht, Germany
Serological pipettes (5 ml, 10 ml, 25 ml)	Th. Geyer Ingredients GmbH & Co. KG, Höxter, Germany
Soft-Ject® 50 ml syringe	Henke-Sass Wolf Mikrooptik GmbH, Nörten-Hardenberg, Germany
Surgical disposable scalpels	Braun Aesculap AG, Tuttlingen, Germany
TC flasks (T_{25} and T_{75})	Sarstedt, Nümbrecht, Germany
TC plate (6-cm and 10-cm)	Greiner Bio-One, Kremsmünster, Austria
TC plate (4-well and 24-well)	Th. Geyer Ingredients GmbH & Co. KG, Höxter, Germany
TC plate (6-well)	STARLAB International GmbH, Hamburg, Germany
Transfection tubes (13 ml)	Sarstedt, Nümbrecht, Germany
Tubes (13 ml)	Sarstedt, Nümbrecht, Germany
UV transparent disposable cuvettes	Sarstedt, Nümbrecht, Germany

4.1.8 Instruments

Table 8: List of instruments used during research.

Instrument		Producer
arium® Lab Water System		Sartorius AG, Göttingen, Germany
Bio-Link 254 UV crosslinker		Vilber Lourmat Deutschland GmbH, Baden-Württemberg, Germany
Centrifuges	Centrifuge 5418	Eppendorf AG, Hamburg, Germany

	Centrifuge Heraeus Fresco21	Thermo Fisher Scientific, Carlsbad, USA
	Centrifuge Heraeus Megafuge 16R	Thermo Fisher Scientific, Carlsbad, USA
	Centrifuge Heraeus Megafuge 1.0R	Thermo Fisher Scientific, Carlsbad, USA
	Centrifuge Heraeus Pico21	Thermo Fisher Scientific, Carlsbad, USA
ChemiDoc™ Touch Imaging System		Bio-Rad Laboratories, Inc., Hercules, USA
FAS V Gel Documentation System (CCD-Sensor)		Nippon Genetics Europe GmbH, Düren, Germany
Gel Electrophoresis Chambers		Thermo Fisher Scientific, Carlsbad, USA
Heidolph magnetic stirrer MR 3000		Merck, Darmstedt, Germany
Heraeus HeraCell 240 incubator		Thermo Fisher Scientific, Carlsbad, USA
Herasafe™ biological safety cabinet		Thermo Fisher Scientific, Carlsbad, USA
Julabo™ shaking water bath SW22		JULABO GmbH, Seelbach, Germany
Microscopes	Olympus BX60 fluorescence microscope	Olympus Corporation, Tokyo, Japan
	Olympus FLUOVIEW FV1000 confocal laser scanning microscope	Olympus Corporation, Tokyo, Japan
	ZEISS Primo Vert	Carl Zeiss Microscopy GmbH, Oberkochen, Germany
Microwave NN-E201WM		Panasonic Corporation, Oaza Kadoma, Japan
Mitsubishi P 95 DE Digitaler Monochrome Printer		Biometra GmbH, Gttingen, Germany
Mr. Frosty™ Cryo Freezing Container		Thermo Fisher Scientific, Carlsbad, USA
NanoDrop™ OneC Spectrophotometer		Thermo Fisher Scientific, Carlsbad, USA
Neubauer improved chamber		Glaswarenfabrik Karl Hecht GmbH & Co. KG, Sondheim/Rhön, Germany
New Brunswick™ Innova® 40/40R Incubator Shaker		Eppendorf AG, Hamburg, Germany
pH meter Hanna Instruments™ HI2211-02		Hanna Instruments Deutschland GmbH, Vöhringen, Germany
PowerPac™ Basic Power Supply		Bio-Rad Laboratories, Inc., Hercules, USA
QuantStudio™ 5 Real-Time PCR System		Thermo Fisher Scientific, Carlsbad, USA
S1000/C1000 Touch Thermal Cyclers		Bio-Rad Laboratories, Inc., Hercules, USA
Safe 2020 Biological Safety Cabinet		Thermo Fisher Scientific, Carlsbad, USA
Scales	Sartorius LC2200P Balance	Sartorius AG, Göttingen, Germany
	Sartorius LC3200D Balance	Sartorius AG, Göttingen, Germany
	VWR SE 1202 Top Load Balance	VWR International, Radnor, USA
Sonifier® S-450A		Branson Ultrasonic Corp., Lawrenceville, USA

Thermomixer 5436	Eppendorf, Hamburg, Germany
Thermomixer comfort	Eppendorf, Hamburg, Germany
Trans-Blot® Turbo™ Transfer System	Bio-Rad Laboratories, Inc., Hercules, USA
Vortex Genie 2™	Bender & Hobein AG, Zurich, Switzerland
VWR MiniStar Silverline Microcentrifuge	VWR International, Radnor, USA
VWR® Tube Rotator	VWR International, Radnor, USA

4.1.9 Buffers and solutions

4.1.9.1 Agarose gel electrophoresis

Table 9: Components of agarose gel used for electrophoresis.

Agarose gel	Agarose 0,5 x TBE 1:20000 GelRed

4.1.9.2 Chemically competent cells

Table 10: List of buffers and their components used for preparing of competent bacterial cells.

Buffer	**Components**
TFB I (1L)	8,1 g $MnCl_2 \cdot 2H_2O$ 7,4 g KCl 1,5 g $CaCl_2 \cdot H_2O$ 3 g KOAc 150 ml glycerol dH_2O pH 6.1
TFB II (1L)	11 g $CaCl_2 \cdot H_2O$ 0,75 g KCl 2,1 g MOPS 150 ml glycerol dH_2O pH 7.0

Buffers TFB I and TFB II were filtered and autoclaved, respectively, and stored at 4°C.

4.1.9.3 Protein extraction

Table 11: List of buffers and their components used for protein extraction.

Buffer	**Components**
Total protein lysis buffer	10 ml Pierce® RIPA Buffer 1x Halt™ Protease Inhibitor Cocktail

	1x tablet of phosphatase inhibitor

4.1.9.4 Western blot

Table 12: List of buffers and their components used for Western blots.

Buffer	Components
Transfer buffer	20% TransBlot Turbo Transfer Buffer (5x) 20% ethanol
TBS (10x)	1,37 M NaCl 100 mM Tris pH 7.6
TBST	10% TBS (10x) 0,1% Tween®20

4.1.9.5 Immunostaining of the cells

Table 13: List of solutions and their components used for ICC.

Solution	Components
Fixation solution	4% Paraformaldehyde
Blocking solution	50mM NH_4Cl in PBS
Permabilization solution	0,2% Triton X-100 in PBS

4.1.9.6 HisPur™ Cobalt Resin purification

Table 14: List of buffers and their components used for recombinant protein purification.

Buffer	Components
Equilibration/Wash Buffer	50mM sodium phosphate 300mM sodium chloride 10mM imidazole pH 7.4
Elution Buffer	50mM sodium phosphate 300mM sodium chloride 150mM imidazole pH 7.4
MES Buffer	20mM 2-(N-morpholine)-ethanesulfonic acid 0.1M sodium chloride pH 5.0

4.1.9.7 Yeast transformation

Table 15: List of solutions and their components used for yeast transformation.

Solution	Components
LiAc mix	1 volume 10xTE pH 7.5 (100 mM Tris-HCl, 10 mM EDTA)

	1 volume 1M lithium acetate 8 volumes autoclaved water
PEG mix	8 volumes 50% PEG 1 volume 10x TE pH 7.5 1 volume 1M lithium acetate

4.1.10 Media

4.1.10.1 Media for bacterial culture

Table 16: List of media and their components used for bacterial culture.

Medium	Components
LB medium	1% Tryptone/Peptone 0,5% Yeast extract 1% NaCl dH$_2$O
2xYT medium	1,6% Tryptone/Peptone 1% Yeast extract 0,5% NaCl dH$_2$O pH 7.0
Ampicillin medium	0,1 mg/ml in LB medium
Kanamycin medium	0,1 mg/ml in LB medium
LB – agar medium	1,5% agar in LB medium
Ampicillin plates	1,5% agar in ampicillin medium
Kanamycin plates	1,5% agar in kanamycin medium

4.1.10.2 Media for cell culture

Table 17: List of media and their components used for cells culture.

Medium	Components
FB medium	DMEM 10% FBS Superior 1% penicillin/streptomycin
mESC medium	DMEM 20% Pansera ES 1% penicillin/streptomycin 0,1 mM NEAA 0,1 mM ß-mercaptoethanol 1000 U/ml LIF
Freezing medium	50% FBS 20% DMSO 30% culture medium

4.1.11 Biological materials

4.1.11.1 Bacterial strains

Table 18: List of bacterial strains used during research.

Strain	Producer
One Shot™ BL21 (DE3) Chemically Competent *E. coli*	Invitrogen™/Thermo Fisher Scientific, Carlsbad, USA
One Shot™ TOP10F' Chemically Competent *E. coli*	Invitrogen™/Thermo Fisher Scientific, Carlsbad, USA

4.1.11.2 Cell lines

Table 19: List of cell lines used during research.

Cell line	Supplier
EDJ #22	ATCC® LGC Standards GmbH, Wesel, Germany
HeLa	ATCC® LGC Standards GmbH, Wesel, Germany
HEK 293T	ATCC® LGC Standards GmbH, Wesel, Germany
NIH 3T3	ATCC® LGC Standards GmbH, Wesel, Germany
Human fibroblasts	Obtained from healthy donors of skin biopsies

4.1.12 Sterilization and autoclaving

Heat-sensitive solutions were filtered using disposable sterile filter units (0.2 to 0.45 µm pore size). All solutions which were not heat-sensitive and the plastic equipment were sterilized at 121°C, 105 Pa for 60 min in an autoclave. Glass wares were sterilized overnight in an oven at 180°C.

4.1.13 Online resources

Table 20: List of online resources and platforms used during research.

Resource	Website
BLAST	https://blast.ncbi.nlm.nih.gov/Blast.cgi
DoubleDigest Calculator	https://www.thermofisher.com/de/de/home/brands/thermo-scientific/molecular-biology/thermo-scientific-restriction-modifying-enzymes/restriction-enzymes-thermo-scientific/double-digest-calculator-thermo-scientific.html
Ensembl	http://www.ensembl.org/index.html
Exac Browser	http://exac.broadinstitute.org
HMMER	https://www.ebi.ac.uk/Tools/hmmer/search/hmmscan

Human Gene Mutation Database (HGMD)	http://www.hgmd.org/
Institute of Human Genetics, UMG, Göttingen	https://www.humangenetik-umg.de
MutationTaster	http://www.mutationtaster.org
Online Mendelian Inheritance in Man® (OMIM®)	https://www.omim.org
PolyPhen-2	http://genetics.bwh.harvard.edu/pph2/
Protein Variation Effect Analyzer (PROVEAN)	http://provean.jcvi.org/index.php
PubMed	https://www.ncbi.nlm.nih.gov/pubmed/
Reverse Complement	https://www.bioinformatics.org/sms/rev_comp.html
Sorting Intolerant From Tolerant (SIFT)	https://sift.bii.a-star.edu.sg/
The Human Protein Atlas	https://www.proteinatlas.org/
United Nations (UN)	https://www.un.org/en/
Varbank	https://varbank.ccg.uni-koeln.de
Webcutter 2.0	http://www.firstmarket.com/cutter/cut2.html
World Health Organization (WHO)	https://www.who.int/

4.1.14 Software

Table 21: List of software used during research.

Software	Application
DNASTAR® FinchTV 1.5.0	Sequences analyzing
EndNote X9	References processing
Image Lab™ Software 6.0	WB images analyzing
Microsoft Excel 2010	Data analyzing
Microsoft PowerPoint 2010	Images preparing
Microsoft Word 2010	Writing
SnapGene Viewer 4.1.6	Plasmids and sequences analyzing

4.2 Methods

4.2.1 Nucleic acids analyses

4.2.1.1 Polymerase chain reaction (PCR)

PCR is the technology which is used to amplify known sequence of DNA. It is based on changing temperature cycles which provide melting and enzymatic replication of DNA. It consists of three major steps: denaturation of DNA double strands, annealing (attachment of primers to complementary fragments of DNA) and elongation of new synthesized strand. Main components which are required to PCR reaction are:

- Buffer which provide stable environment for reaction
- $MgCl_2$ which is required for polymerase activity
- Mix of all nucleotides (dATP, dTTP, dCTP, dGTP) which are used to extend complementary strands (dNTPs)
- pair of primers which initiate reaction in right place
- thermostable polymerase (Taq polymerase) which is major enzyme of reaction

Table 22: Standard PCR mixture components.

Component	Volume
PCR buffer 10x	2,5 µl
50mM $MgCl_2$	0,75 µl
10mM dNTPs	0,5 µl
Forward primer (10 pmol)	0,5 µl
Reverse primer (10 pmol)	0,5 µl
Taq polymerase (5 U/µl)	0,15 µl
DNA	1 µl
H_2O	up to 25 µl

Table 23: Standard PCR program.

Step	Temperature	Time	Number of cycle
Preliminary denaturation	95°C	5 min	
Denaturation	95°C	30 s	
Annealing	58-62°C	30 s	35
Elongation	72°C	30-60 s	
Final elongation	72°C	7 min	

Conditions described above depend on melting temperature (Tm) of primers and size of product.

4.2.1.2 Sequencing PCR

The most frequently used technique of DNA sequencing is the chain-termination method developed by Fredrick Sanger and coworkers in 1977 (Sanger et al., 1977). It provides incorporation of dideoxynucleotides (ddNTPs): ddATP, ddTTP, ddCTP, ddGTP into newly synthesized complementary DNA strand. Each ddNTP is labeled with fluorescence dye and has no 3'-hydroxyl group (3'-OH). 3'-OH group is required for formation of phosphodiester bonds and lack of it results in termination of reaction. This leads to the emergence of mixture of different-length DNA sequences. The sequence is determined based on length of fragments and fluorescent signals.

Prior to sequencing reaction, genomic fragment of interest was subjected to standard PCR reaction described above (see 2.2.1.1 Polymerase chain reaction) using 40 ng od DNA. Subsequently, PCR results were checked on agarose gel and subjected for Exo-SAP purification process. It is a reaction used for enzymatic cleanup of amplified PCR product. It hydrolyzes excess of primers and nucleotides. Purified samples are ready for use in downstream applications such as DNA sequencing.

Table 24: Standard Exo-SAP mixture components.

Component	Volume
SAP	0,3 µl
EXO I	0,075 µl
PCR product	8 µl
H_2O	up to 10 µl

Table 25: Standard Exo-SAP reaction.

Temperature	Time
37°C	20 min
85°C	15 min
10°C	5 min

To perform sequencing PCR Big Dye Terminator V1.1 and V3.1 Cycle Sequencing Kit was used. Reaction was performed using one primer.

Table 26: Standard sequencing mixture components.

Component	Volume
Buffer (5x)	2,25 µl
V3.1	0,25 µl

Primer	0,25 µl
Purified PCR product	0,5 µl
H$_2$O	up to 10 µl

Table 27: Standard sequencing program.

Step	Temperature	Time	Number of cycles
Preliminary denaturation	96°C	30 s	
Denaturation	96°C	10 s	
Annealing	55°C	5 s	40
Elongation	60°C	4 min	
	10°C	5 min	

4.2.1.3 Quantitative Real-Time PCR (qRT-PCR)

This procedure is applied to quantify the levels of gene expression. The assay is rapid and sensitive, and provides detection of amount of PCR product at every cycle of the PCR using fluorescence. QuantiNova™ SYBR® Green PCR Kit was used. The results of qRT-PCR are shown as the number of PCR cycles (Ct – cycle threshold) which are necessary to achieve particular degree of fluorescence (Ponchel et al., 2003).

Table 28: Standard qRT-PCR mixture components.

Component	Volume
SYBR-Green	5 µl
Forward primer	1 µl
Reverse primer	1 µl
cDNA (1:10)	2 µl
H$_2$O	up to 10 µl

Reactions were performed in 384-well PCR plates.

Results were transferred to the Microsoft Excel program. Expression of gene of interest was normalized to housekeeping gene and the relative changes in gene expression were estimated using $2^{-\Delta\Delta Ct}$ method (Livak and Schmittgen, 2001).

4.2.1.4 Isolation of DNA fragments from agarose gels using the QIAquick Gel Extraction Kit (Qiagen)

DNA was extracted from the agarose gel using QIAquick Gel Extraction Kit (Qiagen) following the manufacturer's protocols. . Briefly, gel slices were placed in fresh 1,5 ml tubes and three gel volumes (gel volume corresponds to the slice weight, 100 mg = 100 µl) of buffer QG were added to the agarose gel piece and incubated at 50°C for 10 min.. After gel slice was dissolved 100 µl of isopropanol was added (1 volume) to the sample and vortexed. QIAquick spin columns were placed in 2 ml collection tubes. To bind DNA, the solution was applied to the columns and centrifuged (all centrifugation steps were performed for 1 min at 13000 rpm). Flow-through was discarded and columns were washed with 750 µl of buffer PE. After 5 minutes of incubation, the column was centrifuged, and flow through was discarded. To completely remove residual ethanol, samples were again centrifuged and columns were placed in fresh 1,5 ml tubes. DNA was eluted by application of 30-50 µl of buffer EB to the center of QIAquick membrane, and a subsequent centrifugation. Extracted DNA was stored at -20°C.

4.2.1.5   Isolation of genomic DNA from cells

4.2.1.5.1   DNA extraction using DirectPCR Lysis Reagent

Cells were collected by trypsinization and washed with DPBS. The cell pellets were proceeded immediately or stored at -80°C. The lysis buffer was prepared by adding 5 µl of proteinase K (10 mg/ml) per 100 µl of DirectPCR Lysis Reagent (Tail). The cell pellets were resuspended in 105 µl of lysis buffer and incubated overnight at 55°C with shaking at 600 rpm. The next day, samples were incubated at 85°C for 45 minutes to inactivate proteinase K. To remove the rest of cells the samples were centrifuged for 20 minutes at 14800 rpm (room temperature). Supernatant was transferred to fresh tube and stored at 4°C.

4.2.1.5.2   DNA extraction using NaOH/EDTA solution

Cells were collected by trypsinization and washed with PBS. The cell pellets were proceeded immediately or stored at -80°C. The cell pellets were resuspended in 25 µl of solution containing 25 mM NaOH and 0,25 mM EDTA. Samples were incubated for 30 minutes at 95°C. To stop reaction, 25 µl of 40 mM Tris (pH 7.5) was added. DNA samples were stored at -20°C.

4.2.1.6   Isolation of genomic DNA from tissues using NucleoSpin®Tissue kit

Genomic DNA extraction procedure was performed according to the manufacture protocol. Briefly, small pieces of tissues were placed in 1,5 ml tubes. Lysis buffer was prepared by mixing 180 µl of buffer T1 with 25 µl of proteinase K. 200 µl of lysis buffer was added to samples and vortexed. Samples were incubated overnight at 56°C with 600 rpm shaking. The next day, 200 µl of buffer B3 was added and samples were incubated for 10 minutes at 70°C. Insoluble particles were removed by centrifugation for 5 minutes at 14800 rpm. Supernatant was transferred to fresh 1,5 ml tubes and mixed with 210 µl of 96-100% ethanol. After vortexing samples were applied to the NucleoSpin®Tissue columns and they were centrifuged for 1 minute at 11000 x g. Collection tubes with flow-through were discarded and columns were placed in fresh collection tubes. Columns were washed by adding 500 µl of buffer BW and centrifugation for 1 minute at 1000 x g. Flow-through was discarded, columns were placed back into collection tubes and washed second time with 600 µl of buffer B5 and centrifuged for 1 minute at 11000 x g. To dry membrane, flow-through was discarded and samples were centrifuged for 1 minute at 11000 x g. Columns were placed in fresh 1,5 ml tubes. To elute DNA 100 µl of buffer BE was applied on the membrane, samples were incubated for 1 minute at room temperature and centrifuged for 1 minute at 11000 x g. DNA samples were stored at -20°C.

4.2.1.7 Isolation of total RNA from cells

Cells were collected by trypsinization and washed with PBS. The cell pellets were proceeded immediately or stored at -80°C. The cell pellets were resuspended in 500 µl of Trizol and mixed by pipetting. The samples were incubated for 10 minutes at room temperature and 100 µl of chloroform were added per each 500 µl of Trizol. Tubes were shaken by hands for 15 seconds and incubated 5 minutes at room temperature. Samples were centrifuged at 12000 x g for 15 minutes at 4°C. Aqueous, upper phase was transferred to fresh tube and 250 µl of cold isopropanol supplemented with 1 µl of GlycoBlue were added. Samples were vortexed and incubated at -20°C overnight. The next day samples were centrifuged at 12000 x g for 30 minutes at 4°C. Supernatant was discarded and pellets were washed with 1 ml of 75% ethanol and centrifuged at 12000 x g for 5 minutes at 4°C. Ethanol was removed and pellets were dried at 37°C. RNA pellets were resuspended in 50-100 µl of RNase-free water and mixed by pipetting. RNA samples were stored at -80°C.

4.2.1.8 cDNA synthesis

4.2.1.8.1 Removal of genomic DNA using Amplification Grade DNaseI Kit

1 µg of RNA dissolved in 8 µl nuclease-free water was mixed with 1 µl of 10X Reaction Buffer and 1 µl of DNaseI. Mixture was mixed by pipetting and incubated for 15 minutes at room temperature. To inactivate the DNaseI 1 µl of Stop Solution was added. Samples were incubated for 10 minutes at 70°C to denaturate both the DNaseI and RNA and chilled on ice. Such prepared samples were proceeded with reverse transcription reaction.

4.2.1.8.2 Reverse transcription using RevertAid First Strand cDNA Synthesis Kit

To prepared RNA samples the following components (in the indicated order) were added: 1 µl of Oligo (dT)$_{18}$ primer, 4 µl of Reaction Buffer, 1 µl of RiboLock RNase Inhibitor, 2 µl of dNTP Mix and 1 µl of RevertAid M-MuLV RT. Samples were mixed and centrifuged briefly. Reaction was performed for 1 hour at 42°C and terminated by 5 minute incubation at 70°C. cDNA samples were diluted 1:10 in nuclease-free water and stored at -20°C.

4.2.1.9 DNA cloning

4.2.1.9.1 Amplification of ORFs

Open reading frames (ORFs) of genes of interest were amplified by PCR using specially designed primers. These specific primers allowed flanking amplified ORFs by palindromic sequences recognized by particular restriction enzymes. PCR reaction was followed by agarose gel electrophoresis and gel extraction of DNA of interest.

4.2.1.9.2 Blunt-end cloning

For cloning of blunt-ended PCR products TOPO® Cloning Kits or CloneJET™ PCR Cloning Kit were used.

For TOPO® cloning 4 µl of purified PCR product were mixed with 1 µl of Salt Solution and 1 µl of pCR™4Blunt-TOPO® or pCR™II-Blunt-TOPO® vector and incubated for 30 minutes at 23°C. Reaction was chilled on ice and proceeded with bacterial transformation.

For JET™ cloning 3 µl of purified PCR product were mixed with 5 µl of 2X Reaction Buffer, 1 µl of pJET1.2 vector and 1 µl of T4 DNA Ligase and incubated for 30 minutes at 23°C. Reaction was chilled on ice and proceeded with bacterial transformation.

4.2.1.9.3 TOPO® TA Cloning®

The pCR™2.1-TOPO® plasmid is supplied linearized with single 3´-thymidine (T) overhangs for TA cloning and covalently bound topoisomerase I. Taq polymerase has a nontemplate-dependent terminal transferase activity and adds a single deoxyadenosine (A) to the 3´ ends of PCR product. T overhanging 3' residues allow PCR inserts to ligate efficiently with the vector. For TOPO® TA Cloning® cloning 4 µl of purified PCR product were mixed with 1 µl of Salt Solution and 1 µl of pCR™2.1-TOPO® vector and incubated for 30 minutes at 23°C. Reaction was chilled on ice and proceeded with bacterial transformation.

4.2.1.9.4 The Gateway® Cloning

The Gateway® technology was used to generate expression plasmids that were further used in Bimolecular Fluorescence Complementation (BiFC) assay. The Gateway® technology is based on the bacteriophage lambda site-specific recombination between different attachment sites (*att*) that provides integration of bacteriophage into *E. coli* chromosome. Gene of interest was flanked with specific *attB1* and *attB2* sequences added to 5' ends of forward and reverse primers, respectively. Flanking was performed by PCR reaction. To generate entry clones, BP recombination reaction between *attB*-flanked DNA fragment and *attP*-containing donor vector (pDONR™221) was performed. 3,5 µl of PCR product was mixed with 1 µl of 5x BP Clonase™ II and 0,5 µl of pDONR™221 vector, and incubated overnight at 25°C. To terminate BP reaction, 0,5 µl of Proteinase K was added and mixture was incubated 10 minutes at 37°C. Subsequently, bacterial transformation was performed with BP reaction mixture. Plasmids were extracted from selected colonies with NucleoSpin® Plasmid (NoLid) kit and proceeded with LR recombination reaction. To generate expression clones, LR recombination reaction between *attL*-containing entry clone and *attR*-containing destinations vector (pCSDest C-VC, pCSDest C-VN, pCSDest N-VC, pCSDest N-VN) was performed. 0,5 µl of pDONR™221 entry vector was mixed with 1 µl of 5x LR Clonase™ II, 0,5 µl of destination vector and 3 µl of TE buffer, and incubated overnight at 25°C. To terminate LR reaction, 0,5 µl of Proteinase K was added and mixture was incubated 10 minutes at 37°C. Then, bacterial transformation was performed with LR reaction mixture. Plasmids were extracted from selected colonies with NucleoSpin® Plasmid (NoLid) kit and sequenced. Plasmids with no mutations were next transformed into bacteria and multiplied in bigger volume of LB medium (150 ml). Final expression plasmids were extracted with NucleoBond® Xtra Maxi Plus EF kit and used in BiFC assay.

4.2.1.9.5 Subcloning into expression vectors

Cloning into expression vectors (pCMV-Myc-N, pCMV-HA-N or hEF1α-GFP) was performed by double digestion with particular restriction enzymes on cloning vector containing insert of interest as well as on expression vector. This procedure ensures correct positioning of insert in the expression vector. Digestion was followed by agarose gel electrophoresis and gel extraction. For sticky-end cloning 3 µl of purified insert and 1 µl of purified vector were mixed with 2 µl of 5X Reaction Buffer and 1 µl of T4 DNA Ligase and incubated for 3 hours at room temperature or overnight at 16°C. Reaction was chilled on ice and proceeded with bacterial transformation.

4.2.1.9.6 Preparation of chemically competent *E. coli* Top10 F' cells

E. coli Top10 F' bacteria were cultured overnight in 5 ml of LB medium at 37°C and shaking at 160 rpm. The next day, 900 µl of cultured bacteria were inoculated in 150 ml of fresh LB medium and incubated till OD reach 0,45-0,55. Bacteria were centrifuged at 4°C for 10 minutes at 2000 rpm. Supernatant was discarded and bacterial pellet was resuspended in 30 ml of ice-cold TFBI buffer. Bacteria were incubated for 10 minutes on ice and centrifuged again at 4°C for 10 minutes at 2000 rpm. Supernatant was discarded and bacterial pellet was resuspended in 6 ml ice-cold TFB II. 50 µl aliquots of the bacterial suspension were prepared and immediately frozen in liquid nitrogen. Bacterial aliquots were stored at -80°C until use.

4.2.1.9.7 Transformation of bacteria

Bacterial aliquot and cloning reaction mixture were incubated for 10 minutes on ice to achieve the same temperature. Then cloning mixture was added to the bacteria and mixed gently without pipetting. Bacteria were incubated for 30 minutes on ice and then heat shocked for 45 seconds at 42°C. Bacteria were chilled on ice for 5 minutes and 950 µl of LB medium were added. Bacteria were incubated for at least 1 hour at 37°C with shaking at 600 rpm. To collect bacterial pellets, tubes were centrifuged for 30 seconds at 12000 rpm. Most of supernatant was discarded and pellets were resuspended in remain medium (around 50-60 µl). Bacteria were inoculated on selective plates and incubated overnight at 37°C.

4.2.1.9.8 Culture of bacteria

After selection on plates, positive colonies were picked with toothpick and put into LB medium supplemented with required antibiotic (50µg/ml). For small-scale plasmid purification colonies were incubated in 2,5 ml of medium while for large-scale plasmid purification colonies were incubated in 100 ml of LB medium. Bacteria had access to air and were incubated overnight at 37°C with shaking at 160 rpm.

4.2.1.9.9 Plasmid DNA purification

4.2.1.9.9.1 Small-scale plasmid DNA purification using NucleoSpin® Plasmid (NoLid) kit

Overnight culture was transferred to 1,5 ml tube and centrifuged for 5 minutes at 14800 rpm to collect cells. Bacterial pellets were resuspended in 250 µl of Buffer A1 and vortexed. 250 µl of lysis Buffer A2 were added and samples were mixed by inverting. After 5 minutes incubation at room temperature reaction was stopped by adding 300 µl of Buffer A3 and inverting till samples turned colorless. To clarify lysate, samples were centrifuged for 10 minutes at 11000 x g and supernatant was transferred onto column placed in collection tube. To bind DNA, samples were centrifuged for 1 minute at 11000 x g and supernatant was discarded. Membranes were washed with 600 µl of Buffer A4 and centrifuged for 1 minute at 11000 x g. To dry membranes, supernatant was discarded and samples were centrifuged again for 2 minutes at 11000 x g. Collections tubes with remained buffer were discarded and columns were placed in fresh 1,5 ml tubes. To elute DNA, 50 µl of Buffer AE was added on the membrane. Samples were incubated for 1 minute at room temperature and centrifuged for 1 minute at 11000 x g. Columns were discarded and DNA samples were stored at -20°C.

4.2.1.9.9.2 Large-scale endotoxin-free plasmid DNA purification using NucleoBond® Xtra Maxi Plus EF kit

Overnight culture was transferred to 50 ml falcons and centrifuged for 15 minutes at 4700 rpm to collect cells. Bacterial pellets were resuspended in 12 ml of Buffer RES-EF. 12 ml of Buffer LYS-EF were added and samples were mixed by inverting. For cell lysis, samples were incubated for 5 minutes at room temperature. To stop reaction, 12 ml of Buffer BEU-EF was added and falcons were inverted till samples turned colorless. Samples were incubated for 5 minutes on ice. Meantime, NucleoBond® Xtra Columns with filters were equilibrated with 35 ml of Buffer EQU-EF and emptied by gravity flow. Lysates were loaded to the equilibrated columns and columns were left to empty by gravity flow. Columns with filters were washed with 10 ml of Buffer FIL-EF, emptied with gravity flow and filters were discarded. Subsequently, columns were washed with 90 ml of Buffer ENDO-EF and 45 ml of Buffer WASH-EF. After each washing columns were emptied by gravity flow. To elute DNA, columns were washed with 15 ml of Buffer ELU-EF and the elution fraction was collected in 50 ml falcon. DNA was precipitated by adding 10,5 ml of isopropanol, vortexing and incubation for 2 minutes at room temperature. Precipitated DNA was loaded into 30 ml syringe with attached NucleoBond® Finalizer. Sample was loaded in finalizer by slowly pressing through the finalizer. Finalizer were washed with 5 ml of 70% ethanol in the same way. Finalizers were dried by pressing air through them. To elute DNA, 500 µl of Buffer TE-EF was slowly pressed through the finalizer and collected in fresh 1,5 ml tube. DNA samples were stored at -20°C.

4.2.1.10 Measurement of concentration by NanoDrop™ OneC Spectrophotometer

NanoDrop™ measurement of concentration is based on spectrophotometry which is the quantitative measurement of the reflection or transmission properties of a material as a function of wavelength.

4.2.1.10.1 Nucleic acids concentration

Nucleic acids (DNA and RNA) concentrations were measured using measurement pedestal. Concentration was based on absorbance at 260 nm and the defined extinction coefficient. Blanking was performed using 1,5 µl of water in which the nucleic acids were dissolved. Then 1,5 µl of DNA or RNA was put on pedestal and measured. The program estimated the nucleic acid (c) concentration using following equation:

$$c = \frac{A \times \varepsilon}{b}$$

c – the nucleic acid concentration in ng/µl

A – the absorbance in AU

ε – the wavelength-dependent extinction coefficient in ng-cm/µl

b – the path length in cm

To check the quality of nucleic acid the ratio of absorbance in different wavelength was estimated. A ratio 260/280 of ~1.8 indicates good quality of DNA while a ratio of ~2.0 indicates "pure" RNA. If the ratio is appreciably lower in either case, it may indicate the presence of protein, phenol or other contaminants that absorb strongly at or near 280 nm. The 260/230 ratio is a secondary measure of nucleic acid purity. It is commonly in the range of 1.8-2.2. If the ratio is appreciably lower, this may indicate the presence of co-purified contaminants.

4.2.1.10.2 OD600 of bacterial culture

To monitor growth rate of bacterial cultures, the optical density (OD) of the culture in growth media was measured. The OD600 application measures light transmission and uses that value to calculate absorbance at 600 nm wavelength. The 1 cm plastic cuvettes were used for OD600 measurements.

4.2.2   Cell culture

All procedures with cell cultures were performed in sterile conditions under cabinet sterile with UV light. All media and other materials were disinfected with 70% ethanol before put them under cabinet. Media were also heated for at least 30 minutes in water bath at 37°C. Cells were cultured in flasks (T_{25} or T_{75}) and on plates (24-well, 6-well, 6 cm, 10 cm) under conducive conditions in incubator (37°C, 5% CO_2, and 95% of humidity). Media were changed daily or every second day. Each time cells were washed with PBS.

4.2.2.1 Subculture

To avoid overgrowth of cultures, cells were passaged when confluence was near 90-100%. Cells were washed with DPBS and incubated in minimal amount of trypsin at 37°C until the cells detached from the bottom of the culture flask (checked under microscope). Then equal volume of medium was added to inactivate enzyme. Cell suspension was transferred to 15 ml falcon and then centrifuged for 5 minutes at 1000 rpm. Supernatant was sucked out and cells were resuspended in fresh medium, and put into new flasks or on new plates.

4.2.2.2 Counting

To obtained particular number of cells in culture, the counting of cells in was performed. 10 µl of cell suspension was loaded onto Neubauer chamber. Cells were counted in four squares of chamber and then the number of cells was estimated with use of equation:

$$\frac{x}{4} \times 10^4 = number\ of\ cells/ml$$

4.2.2.3 Cryopreservation

Cells were grown to a confluency of > 80%, washed with DPBS, and trypsinised. Cell suspension was transferred to 15 ml falcon and then centrifuged for 5 minutes at 1000 rpm. Supernatant was sucked out and cells were resuspended in fresh medium. The same volume of freezing medium was added and cell suspension was aliquoted to cryotubes (1 ml in each). Cryotubes were well closed and placed in freezing container. Cells' stocks were stored in freezing container for 24 hours at -80°C and transferred for longer storage to -150°C.

4.2.2.4 Feeder layer preparation

Inactivated mouse embryonic fibroblasts (MEFs) were used as a feeder layer for embryonic stem (ES) cells culture. MEF was cultured under normal conditions till 24 T75 flasks were fully confluent. Then, cells were incubated with Mitomycin C (10 µg/ml) for 3 hours under normal culture conditions. Cells were washed three times with PBS and 72 stocks were frozen and stored at -150°C.

4.2.2.5 Transfection of cells

Transfection was performed with Lipofectamine™ 2000 Reagent. 24 hours before transfection appropriate number of cells were sowed on plates (see Table 29). Transfection complex was prepared according manufacture protocol (see Table 29). Briefly, nucleic acid solution and Lipofectamine™ 2000 solution were mixed with Opti-MEM® I Reduced Serum Medium in two separate transfection tubes. Both solutions were mixed gently by shaking and incubated for 5 minutes at room temperature. Nucleic acid solution was added to the Lipofectamine™ 2000 solution and mixed gently by shaking. Transfection mixture was incubated for 20 minutes at room temperature to obtain nucleic acid-Lipofectamine™ 2000 complexes. Meantime, culture medium was exchanged to Opti-MEM® (heated to 37°C). Transfection complex was added to cells (see Table 29) and plates were incubated for 4 hours at 37°C. Transfection solution was exchanged to normal culture medium. 24-48 hours after transfection cells were proceeded for other experiments.

Table 29: Standard Lipofectamine™ 2000 transfection mixtures and their components.

DNA transfection					
culture plate	number of cells	DNA	Lipofectamine™ 2000	Opti-MEM® used for solutions	Opti-MEM® used for cultures
6-well	5×10^5	4,0 µg	10 µl	250 µl	2 ml
24-well	5×10^4	0,8 µg	2 µl	50 µl	500 µl
siRNA transfection					
culture plate	number of cells	RNA	Lipofectamine™ 2000	Opti-MEM® used for solutions	Opti-MEM® used for cultures
6-well	3×10^5	100 pmoli	5 µl	250 µl	2 ml

4.2.2.6 Immunostaining of cells

For staining, cells were culture in 24-well plates on glass cover slips. Before staining, cells were counted and seated on plates and/or transfected. Culture medium was removed from wells and cells were washed two times with pre-warmed PBS (37°C). Fixation was performed using of 4% paraformaldehyd (PFA) in PBS (pre-warmed at 37°C) for 30 minutes at room temperature. Cells were washed again two times with PBS (from this step room temperature) and blocked with 50mM NH_4Cl for 10 minutes. Then two washings in PBS were performed. Permabilization of cell membrane was done by incubation three times for 4 minutes with 0,2% Triton X-100 in PBS. Then cells were incubated with primary antibody (diluted 1:100 in 0,2% Triton X-100) for 1 hour at room temperature in staining chamber. Cells were washed with 0,2% Triton X-100 three times for 4 minutes and incubated with secondary antibody (diluted 1:300 in 0,2% Triton X-100) for 1 hour at room temperature in staining chamber. Afterwards the cells were washed with 0,2% Triton X-100 three times for 4 minutes and two times with PBS for 5 minutes (all washings were performed in the dark to avoid loss of fluorescent signal). ProLong™ Diamond Antifade Mountant with DAPI was dropped onto a slide and covered by coverslip with cells. Slides were left for 24 hours at room temperature to allow DAPI mountant medium to polymerize. Slides were stored at 4°C.

4.2.2.7 Bimolecular Fluorescence Complementation (BiFC) assay

The Gateway® cloning resulted in generation of numerous expression plasmids with different combinations of fused proteins:

- pCSDest C-VC – C-terminal part of fluorescent Venus protein linked to the C-terminus of protein of interest
- pCSDest C-VN – N-terminal part of fluorescent Venus protein linked to the C-terminus of protein of interest
- pCSDest N-VC – C-terminal part of fluorescent Venus protein linked to the N-terminus of protein of interest
- pCSDest N-NC – N-terminal part of fluorescent Venus protein linked to the N-terminus of protein of interest

Subsequently HeLa cells were co-transfected with NARF, NARFp.H367R, LMNA, CBX5 in all possible combinations to evaluate direct interactions between mentioned proteins. Subsequently, cells were fixed with 4% PFA 24 hours after transfection. ProLong™ Diamond Antifade Mountant with DAPI was dropped onto a slide and covered by coverslip with cells. Slides were left for 24 hours at room temperature to allow DAPI mountant medium to polymerize. Slides were stored at 4°C.

4.2.2.8 Stress experiment

24 hours before experiment, cells were counted and 4×10^5 cells were seated on 6 cm plate. The next day cells were treated with etoposide or UV light. Cells treated with etoposide were washed with 2 ml of PBS and incubated with 50 μM etoposide in 3 ml of normal culture medium for 1 hour under conducive conditions in incubator. Afterwards, cells were washed with 2 ml of PBS and medium was changed to normal culture medium. Cells treated with UV light were placed in the UV crosslinker device without lids and treated with UV-C radiation with an energy of 10 mJ/cm^2. Cells were washed with 2 ml of PBS and medium was changed to normal culture medium. Cells for protein analysis were harvested by trypsinization 1, 6 and 24 hours after treatment. Cells pellet were proceeded for total protein extraction immediately or stored at -80°C. Cells for staining were fixed with 4% paraformaldehyde 1, 6 and 24 hours after treatment and proceeded with immunostaining immediately or stored in PBS at 4°C. For both analyses proper negative, untreated controls were prepared.

4.2.2.9 Oxidative stress experiment

Staining procedure was performed using CellROX® Deep Red Flow Cytometry Assay Kit. Cells were collected and the concentration of cells was adjusted to 5×10^5 cells/ml of growth medium. For each cell line appropriate positive and negative controls were prepared (see Table 30).

Table 30: List of samples prepared for oxidative stress experiment.

Sample	ROS induction	Antioxidant treatment	CellROX® Deep Red staining	SYTOX® Blue Dead Cell staining
Untreated, unstained control	–	–	–	–
Untreated, double stained control	–	–	+	+
Negative control, double stained	+	+	+	+
Positive control CellROX® Deep Red staining	+	–	+	–
Positive control SYTOX® Blue Dead Cell staining	+	–	–	+
Positive control, double stained	+	–	+	+

Reactive oxygen species (ROS) generation in cells was induced with tert-butyl hydroperoxide (TBHP) treatment. Cells were incubated with 200 µM TBHP for 1 hour at 37 °C. The negative control was treated with antioxidant before ROS production induction. Cells were incubated with 250-1000 µM N-acetylcysteine (NAC) for 1 hour at 37 °C before THBP treatment. To visualize ROS production in cells, they were subsequently stained with 500 nM CellROX® Deep Red for 1 hour at 37 °C. To visualize dead cells in suspension 1 µM of SYTOX® Blue Dead Cell stain was added at the final 15 minutes of staining. The samples were analyzed with flow cytometry, using 405-nm and 635-nm excitation for SYTOX® Blue Dead Cell stain and CellROX® Deep Red reagent, respectively. Fluorescence emission was analyzed with 450/50 BP and 665/40 BP filters for SYTOX® Blue Dead Cell stain and CellROX® Deep Red detection reagent, respectively. Untreated, unstained controls were used to adjust forward versus side scatter (FSC vs SSC) gating to identify cells of interest. Positive single stained controls were used as single-color compensation controls.

4.2.2.10 Proliferation assay

Proliferation assay was performed using CellTiter 96® AQ$_{ueous}$ One Solution Reagent. It is a colorimetric method that allows for determining the number of viable cells. CellTiter 96® AQ$_{ueous}$ One Solution Reagent contains the MTS tetrazolium compound which is bioreduced by cells, and phenazine ethosulfate (PES) that is electron coupling reagent enhancing chemical stability. Cells reduce MTS tetrazolium into colored formazan product which quantity can be measured by absorbance. Cells were collected and seeded in 24-well plates, at the number of 1,5 x 10^4, in six repeats. Next day, 100 µl of CellTiter 96® AQ$_{ueous}$ One Solution Reagent was added to each well and cells were incubated for 3 hours under normal culture conditions. The quantity of formazan product reflecting the number of living cells was estimated by measuring of absorbance at 490 nm with NanoDrop™ OneC Spectrophotometer. The measurement was repeated after additional 24 hours and relative change of living cells number was estimated by relative change in absorbance.

4.2.3   Protein analyses

4.2.3.1   Protein Extraction

4.2.3.1.1   Protein fractionation using NE-PER Nuclear and Cytoplasmic Extraction Reagents

Cells were collected with trypsinization and pellets were washed with PBS. For nuclear and cytoplasmic fractionation cells' pellets were proceeded immediately and the volume ratio of reagents Cytoplasmic Extraction Reagent I (CER I):Cytoplasmic Extraction Reagent II (CER II):Nuclear Extraction Reagent (NER) was maintained according to manufacturer's protocol. Reagents CER I and NER were pre-mixed with protease inhibitor cocktail. Cells' pellets were resuspended in ice-cold CER I by vigorous vortexing for 15 seconds and incubated 10 minutes on ice. Ice-cold CER II was added and samples were vortexed for 5 seconds, incubated for 1 minute on ice and vortexed for 5 seconds again. Samples were centrifuged at 4°C for 5 minutes at 14800 rpm and cytoplasmic extract was transferred to fresh pre-chilled tube. Remained pellets were resuspended in ice-cold NER. Samples were vortexed for 15 seconds and incubated for 40 minutes on ice with 15 seconds vortexing every 10 minutes. Samples were centrifuged at 4°C for 10 minutes at 14800 rpm and nuclear extract was transferred to fresh pre-chilled tube. Depending on the type of experiment, the remaining pellets were discarded or subjected to extraction of remained insoluble proteins. Protein extracts were stored at -80°C.

4.2.3.1.2 Total protein extraction

Cells were collected by trypsinization and washed with PBS. The cells' pellet were proceeded immediately or stored at -80°C. The cell pellets were suspended in 200 µl of Pierce™ RIPA Buffer containing protease and phosphatase inhibitors. Cell suspension was sonicated three times for 10 periods using microtip, power output 3 and duty cycle 30%. Samples were centrifuged at 4°C for 20 minutes at 13000 rpm. Supernatant was transferred to fresh pre-chilled tube. Protein extracts were stored at -80°C.

4.2.3.2 Measurement of protein concentration using Pierce™ BCA Protein Assay Kit

This assay allows for the colorimetric detection and quantitation of total protein. To prepare samples for measuring 10 µl of protein were mixed with 200 µl of BCA Working Reagent (WR). WR was previously prepared by mixing 50 parts of BCA Reagent A with 1 part of BCA Reagent B. 200 µl of WR without protein was used as blank sample. Samples were mixed and incubated for 30 minutes at 37°C. Subsequently, samples were transferred to special plastic cuvettes and absorbance was measured at 562 nm wavelength using NanoDrop™ One^C. Protein concentration was estimated based on a previously plotted standard curve defined by known albumin (BSA) concentrations.

4.2.3.3 Western blot

4.2.3.3.1 SDS-PAGE electrophoresis

To prepare samples required amount of protein, they were mixed with water to obtain the equal volume of all samples. Samples were mixed with 4X Laemmli Sample Buffer and 10X Reducing Agent (RA), and incubated for 5 minutes at 95°C in order to denaturate proteins. Samples were chilled on ice for 5 minutes and centrifuged briefly. Meantime, the gel in the chamber was prepared according manufactural instruction. Samples and 10 µl of Precision Plus Protein™ All Blue Standards were loaded onto gel. Electrophoresis was conducted at 80V till all samples leave wells and then at 120V until the dye reaches the reference line on the bottom of cassettes. Gel was removed from cassette and prepared for transfer. Before transfer, gel picture of proteins was taken.

4.2.3.3.2 Transfer on membrane

Transfer of proteins was performed using Trans-Blot® Turbo™ Transfer System and RTA Transfer Kit (Mini, PVDF). For activation PVDF membrane was immersed in 100% methanol, washed with water and then equilibrated for 2-3 minutes in cold transfer buffer. Transfer stacks were immersed in transfer buffer as well. Next all components were placed in the cassette as a "sandwich" in the following order: bottom wetted stack, equilibrated membrane, gel, top wetted stack. Air bubbles and excess buffer were removed with roller. Cassette was locked and inserted in the instrument. Transfer was conducted with one of Bio-Rad preprogrammed protocols – MIXED MW (1,3 A constant for 1 Mini gel or 2,5 A constant for 2 Mini gels; up to 25 V; 7 minutes). Membranes were immediately proceeded to probing and developing.

4.2.3.3.3 Probing and developing

To avoid unspecific binding of antibodies, membrane was blocked by incubation with 5% milk in TBST for 1 hour. Membrane was probed with primary antibody diluted 1:1000 in 2% milk in TBST overnight at 4°C, and additional 1 hour at room temperature. Membrane was washed three times for 20 minutes with 2% milk in TBST, and then incubated with secondary antibody diluted 1:10000 in 2% milk in TBST for 1 hour at room temperature. To remove unbound antibodies next series of washing was applied – three times for 20 minutes in 2% milk in TBST and three times for 10 minutes in PBS. To detect signal Clarity™ Western ECL Substrates and ChemiDoc™ Touch Imaging System were used.

4.2.3.3.4 Probing and developing – His-tag

This protocol was used for detection of recombinant His-tagged proteins. After transfer, membrane was washed two times for 10 minutes with TBS and blocked with 3% BSA in TBS for 1 hour at room temperature. Membrane was washed two times for 10 minutes with 0,05% Tween 20 and 0,2% Triton X-100 in TBS, and one time for 10 minutes with TBS. Membrane was probed with His-tag antibody diluted 1:1000 in 3% BSA in TBS overnight at 4°C, and additional 1 hour at room temperature. Membrane was washed two times for 10 minutes with 0,05% Tween 20 and 0,2% Triton X-100 in TBS, and one time for 10 minutes with TBS. Membrane was incubated with secondary antibody diluted 1:10000 in 2% milk in TBST for 1 hour at room temperature. To remove unbound antibodies next series of washing was applied – three times for 20 minutes in 2% milk in TBST and three times for 10 minutes in PBS. To detect signal Clarity™ Western ECL Substrates and ChemiDoc™ Touch Imaging System were used.

4.2.3.3.5 Stripping and reprobing

To reprobe membrane with different antibodies, first antibody was removed from membrane using Restore™ Plus Western Blot Stripping Buffer. After developing, membrane was washed with PBS, and incubated with stripping buffer for 15 minutes at room temperature.

Then membrane was washed with TBST and blocked with 5% milk in TBST for 1 hour at room temperature. Membrane was reprobed with primary antibody and all steps were performed in the same way as for the first development.

4.2.3.4 Co-immunoprecipitation using Immunoprecipitation Kit Dynabeads® Protein G

Co-immunoprecipitation (CoIP) method was used to study interaction between two proteins. For each reaction 50 µl of Dynabeads® was used. Beads were properly resuspended, transferred to fresh 1,5 ml tube and supernatant was removed using magnet. To bind antibody, beads were incubated with 10 µg of antibody diluted in 200 µl of Ab Binding & Washing Buffer. Incubation was carried out with rotation for 10 minutes at room temperature, and then overnight at 4°C. Antibody solution was removed and beads were washed once with 200 µl of Ab Binding & Washing Buffer. Beads-antibody complexes were incubated with protein extracts obtained from cells (10 µl of extracts were stored at -80°C as input for further analysis). Incubation was carried out with rotation for 10 minutes at room temperature, and then overnight at 4°C. Supernatant was transferred to fresh tube and stored at -80°C for further analysis. Beads were washed three times with 200 µl of Washing Buffer, resuspended in 100 µl of Washing Buffer and transferred to fresh tube to avoid elution of proteins attached to tube wall. Supernatant was removed and beads were resuspended in 20 µl of Elution Buffer and 10,5 µl of pre-mixed Laemmli Sample Buffer and Reducing Agent (prepared according manufacturer's instruction). Beads-antibody-proteins complexes were eluted and denatured for 10 minutes at 70°C. CoIP elutions and inputs were loaded on gel and proceeded with Western Blot protocol.

4.2.3.5 Pull-down assay

For poorly expressed and hardly soluble proteins for which CoIP was not possible, a pull-down interaction analysis using recombinant proteins was performed.

4.2.3.5.1 Preparing of chemically competent *E. coli* BL21 Star™ (DE3) One Shot® cells

Chemically competent cells were prepared as previously described (see 2.2.1.9.6 Preparation of chemically competent E. coli Top10 F' cells).

4.2.3.5.2 Recombinant protein expression in BL21 Star™ (DE3) One Shot® bacteria

For production of recombinant proteins, ORFs of genes of interests were cloned into pET28 a(+) vector. DNA was introduced into bacterial cells with basic transformation procedure. Briefly, bacterial stock was thawed on ice and 10 ng of plasmid DNA was added and mixed without pipetting. Bacteria were incubated for 30 minutes on ice and then heat shocked for 30 seconds at 42°C. Bacteria were chilled on ice for 5 minutes and 250 µl of LB medium were added. Bacteria were incubated for 1 hour at 37°C with shaking at 600 rpm. 60 µl of bacterial suspension was inoculated on LB plates containing ampicillin and incubated overnight at 37°C. One colony was picked with toothpick and incubated in 10 ml of 2x YT medium containing ampicillin overnight at 37°C with shaking at 160 rpm. The next day, pre-culture was inoculated in 500 ml of fresh 2x YT medium containing ampicillin and incubated at 37°C with shaking at 160 rpm until OD_{600} reached ~0,4. Bacterial cultures were induced by adding IPTG to a final concentration of 0,5 mM and incubation for 3 hours at room temperature with shaking. Induced bacteria were collected by centrifugation at 4°C for 10 minutes at 4700 rpm. To release proteins, bacterial pellets were resuspended in 10 ml of Equilibration/Wash Buffer (see 1.2.12.5.3 Purification of recombinant proteins using HisPur™ Cobalt Resin) containing protease inhibitor cocktail and sonicated three times for 1 minute on ice using flat tip, power output 8 and duty cycle 30%. Samples were centrifuged at 4°C for 20 minutes at 14800 rpm and supernatant was collected in fresh 15 ml falcon and stored at -80°C and proceeded with HisPur™ Cobalt Resin purification.

4.2.3.5.3 Purification of recombinant proteins using HisPur™ Cobalt Resin

To purify His-tagged proteins expressed in BL21 Star™ (DE3) One Shot® bacteria procedure using gravity-flow column and native conditions was used. Columns were packed with 1 ml of cobalt resin and left to remove storage buffer by gravity flow. Columns were equilibrated with 2 ml of Equilibration/Wash Buffer. 10 ml of bacterial protein extract in Equilibration/Wash Buffer was loaded onto the resin and the flow-through was collected in fresh 15 ml falcon. Resin was washed two times with 2 ml of Equilibration/Wash Buffer. To elute His-tagged proteins, the resin was washed with 4 ml of Elution Buffer and flow-through was collected in fresh 1,5 ml tubes – 500 µl each (elution 1-8). All elution samples were checked by Western blot. For further analysis proteins were concentrated in smaller amount using Centrisart® I (20,000 MWCO CTA) ultrafiltration spin columns. Centrifugation was performed few times at 4°C for 10 minutes at 2000 rpm till volume reached about 500 µl. Protein extracts were stored at -80°C. The cobalt resins were reused several times. For that purpose, they were regenerated by washing with 10 ml of MES Buffer and 10 ml of ultrapure water. The resins were stored as 50% slurry in 20% ethanol at 4°C.

4.2.3.5.4 Pull-down

The concentrated His-tagged recombinant proteins were mixed with nuclear or cytoplasmic protein fraction extracted from transfected or untransfected cells. Mixture was incubated overnight at 4°C on roller. His-tagged proteins together with bounded interaction partners were purified on HisPur™ Cobalt Resin using the same gravity-flow and native conditions procedure. 10-20 µl of each protein extract were left as input control. Pull-down of interacting proteins was analyzed by Western blot.

4.2.4 Knock-in mouse model generation

4.2.4.1 Preparation of cells for blastocyst injection

4-5 days prior to injection cells were plated onto gelatinized T_{75} flasks containing feeder layer (mitomycin C-treated MEFs) to obtain appropriate number of cells. 1-2 days prior to injection cells were passaged to get consistent density of dividing cells. On the day of injection, 3-4 hours before trypsinization, the culture medium was changed. Cells were washed with PBS and trypsin was added. Trypsinization was performed 10-15 minutes at 37°C to obtained single cell suspension (checked under microscope). Equal volume of medium was added to inactivate trypsin and cells were pipetted up and down to disrupt clumps. Next, cells were transferred to 15 ml falcons and centrifuged for 5 minutes at 1000 rpm. Cells were resuspended in 10 ml of ES medium and seeded on gelatinized 10 cm plates for 20 minutes to remove feeder cells. Then, cells were collected again in 15 ml and centrifuged for 5 minutes at 1000 rpm. Cells pellet was resuspended in 1 ml of ES medium containing 20 mM of HEPES. Cells were transferred on ice to Max Planck Institute of Experimental Medicine (Göttingen), where procedure of blastocyst injection was performed.

4.2.5 Yeast complementation experiments

All yeast experiments were performed in collaboration with Prof. Blanche Schwappach lab at the Department of Molecular Biology (UMG). All materials and instruments were provided by Prof. Schwappach lab, and all procedures were performed with kind help of Àkos Farkas and Ariane Wolf.

4.2.5.1 Yeast transformation (quick and dirty, PEG-LiAc method)

Day before transformation, yeast strains were inoculated for overnight culture to obtain cells in logarithmic phase and increase transformation efficiency. Cells were centrifuged for 1 minute at 1500 x g and resuspended in 120 µl of LiAc mix. Next, 200-300 ng of plasmid DNA, 14 µl of carrier DNA and 750 µl of PEG mix were added. Mixture was mixed and yeasts were incubated 1 hour at 42°C. Cells were centrifuged for 1 minute at 3000 rpm, and supernatant was discarded. Cells were resuspended in 50-100 µl of autoclaved water. Subsequently, yeasts were inoculated on selective plates in decreasing concentrations. Plates were incubated at least 2 days in different temperatures, and the ability of protein of interest to complement was analyzed by presence or lack of growing yeast cells.

4.2.6 Statistics analyzes

Generated data were analyzed using Microsoft Excel. The two-tailed student's t test was used to evaluate differences between samples. Significances are indicated: $*p<0.05$, $**p<0.01$.

5 Results

5.1 Identification of novel progeria-associated gene

Recently, our group studied a 4½-year-old girl diagnosed with a congenital, segmental progeria syndrome. She presented with developmental delay, intellectual disability, mild primary microcephaly, short stature, dilated cardiomyopathy, reduced subcutaneous fat, wrinkly skin and sparse hair, and a progeroid facial expression. She showed no recurrent infections or any other signs of immunological problems. The clinical observations were in accordance with the diagnosis of a progeroid syndrome, with an early onset in the first years of life, partially resembling the Cockayne syndrome. We performed trio whole-exome sequencing (WES) on DNA extracted from blood lymphocytes of the patient and her parents. Using an in-house analysis pipeline established by our research group, we conducted WES data analysis and in addition filtering of variants with the Varbank platform (varbank.ccg.uni-koeln.de). We applied the following criteria for the filtering of the WES variants: coverage of over ten reads, a minimum quality score of 10, allele frequency $\geq$ 20%, and a minor allele frequency (MAF) < 0.000001 in the ExAC database. Using these filter criteria, we found that WES revealed only one variant, which was present in the patient but not in both parents. This variant was located in the *Nuclear Prelamin A Recognition Factor* (*NARF*) gene, and *de novo* occurrence was confirmed through Sanger sequencing (Figure 5a). The c.1100A>G mutation converts a highly conserved histidine at position 367 to arginine (p.His367Arg; Figure 5b) and is predicted to be damaging by several prediction programs, including MutationTaster, PolyPhen2, PROVEAN, and SIFT. At the start of my thesis, NARF function was only poorly characterised. NARF was shown to interact with pre-lamin A in a farnesylated manner (Barton and Worman, 1999). As mutations in *Lamin A* (*LMNA*) lead to HGPS, the most prominent premature ageing disorder, we considered the p.His367Arg mutation in *NARF* highly relevant, and we considered *NARF* a promising candidate gene regarding the observed phenotype in the patient (Schotik, 2017). In addition to this hypothesis, patient-derived fibroblasts exhibited an abnormal nuclear structure presented by increased blebbing of the nuclear lamina (data not shown; Schotik, 2017). Interestingly, in *C. elegans*'s mutant *oxy-4*, a missense mutation in the identical region induced increased sensitivity to oxidative stress and decreased lifespan of worms (Figure 5c; Fujii et al., 2009).

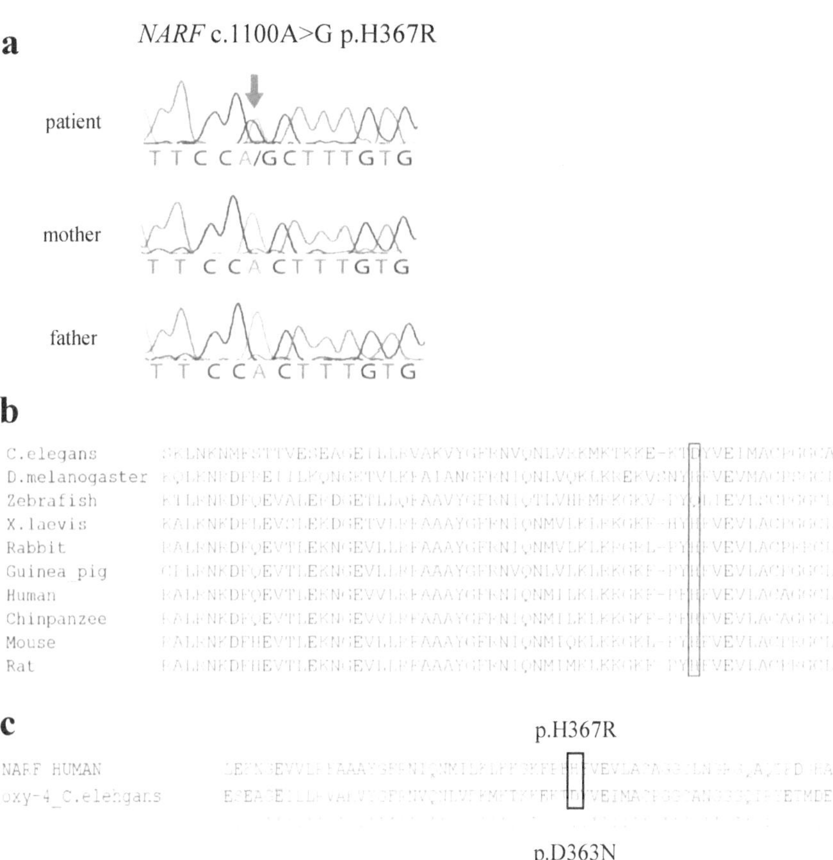

Figure 5: Identification of a *de novo* mutation in *NARF* in a patient with a progeroid syndrome phenotype.
(a) Chromatograms illustrating the *de novo* mutation c.1100G>A in *NARF*. (b) Cross-species alignment depicting conservation of the p.H367 residue (boxed in black) affected by the identified c.1100A>G mutation. (c) Cross-species alignment illustrating conserved region of mutation (boxed in black) in human and *C. elegans*.

5.2 Anti-NARF/Narf antibody generation

Antibodies are essential research tools that have enabled researchers from various fields of science to identify, locate, and quantify protein targets. Since commercially available anti-NARF antibodies were unspecific or inefficient, I decided to generate a new antibody recognising both human and mouse NARF/Narf proteins for further functional *in vitro* and *in vivo* studies. I developed an anti-NARF/Narf (further referred to as anti-NARF) antibody in collaboration with the Eurogentec company. The strategy was based on the immunisation of animals with two different 16-amino-acid peptides (peptide A and peptide B), which were common for mouse and human proteins without any homology for other proteins. To increase the chance of generating a specific antibody, we used two host species: rabbit and llama. We immunised two rabbits and two llamas. We injected one animal in each pair with peptide A and the other with peptide B, culminating in the generation of four anti-NARF antibodies. I further tested the specificity of the generated antibodies on protein extracts from HeLa cells. Protein extracts from *NARF*-overexpressing HeLa cells served as a control. To analyse the subcellular localisation of NARF, I split the protein extracts into cytosolic and nuclear fractions. Western blot (WB) analysis revealed that the peptide-A-derived antibodies, both rabbit and llama, were unspecific, resulting in the appearance of several bands (data not shown). We thus excluded these two antibodies from further studies. By contrast, the peptide-B-derived antibodies showed specificity for two proteins: the exogenous overexpressed Myc-NARF protein (upper band ~52 kDa) and a smaller protein (lower band ~42 kDa) that was present in both, the cytoplasmic and nuclear protein fractions, independent of NARF transfection, showing the endogenous origin of this protein (Figure 6a). To test whether the lower band corresponded to endogenous NARF, I performed a knock-down (KD) assay using small interfering ribonucleic acid (siRNA) transfection in the HeLa and HEK293T cells. Quantitative real-time PCR (qRT-PCR) analysis demonstrated that siRNA transfection with two different siRNAs or a combination thereof reduced the expression of the *NARF* gene to ~10-20% and ~40-60% in the HeLa and HEK293T cell lines, respectively (Figure 6b). Transfection with scrambled siRNAs (Ctrl) served as a control. WB analysis performed on the total proteins extracted from both cell lines showed no difference between the NARF siRNA and Ctrl siRNA treated cells in terms of the 42 kDa band intensity (Figure 6c), which did not decrease, indicating that the recognised protein was not the endogenous NARF. On testing proteins extracted from WT and *Narf* KD mouse NIH 3T3 cells, I obtained similar results (data not shown). To further characterise the affinity of the produced antibodies, I performed immune precipitation (IP) assays in HeLa cells, using the rabbit antibody, followed by proteomic analysis. Rabbit IgG IP served as a negative control (Figure 6d). I used the gel fragment containing the proteins separated with SDS-PAGE (proteins corresponding to 28-62kDa; Figure 6d, red frame) for mass spectrometry analysis performed by the Proteomics Service Facility, Göttingen. Proteomics analysis revealed that the generated antibody exhibited high affinity to β-actin (1030 reads) but only minor affinity to endogenous NARF protein (4 reads; Figure 6e). Taken together, the generated NARF antibody was able to detect

NARF (~52 kDa), but it exhibited a much higher affinity to β-actin in the protein extracts from HeLa cells.

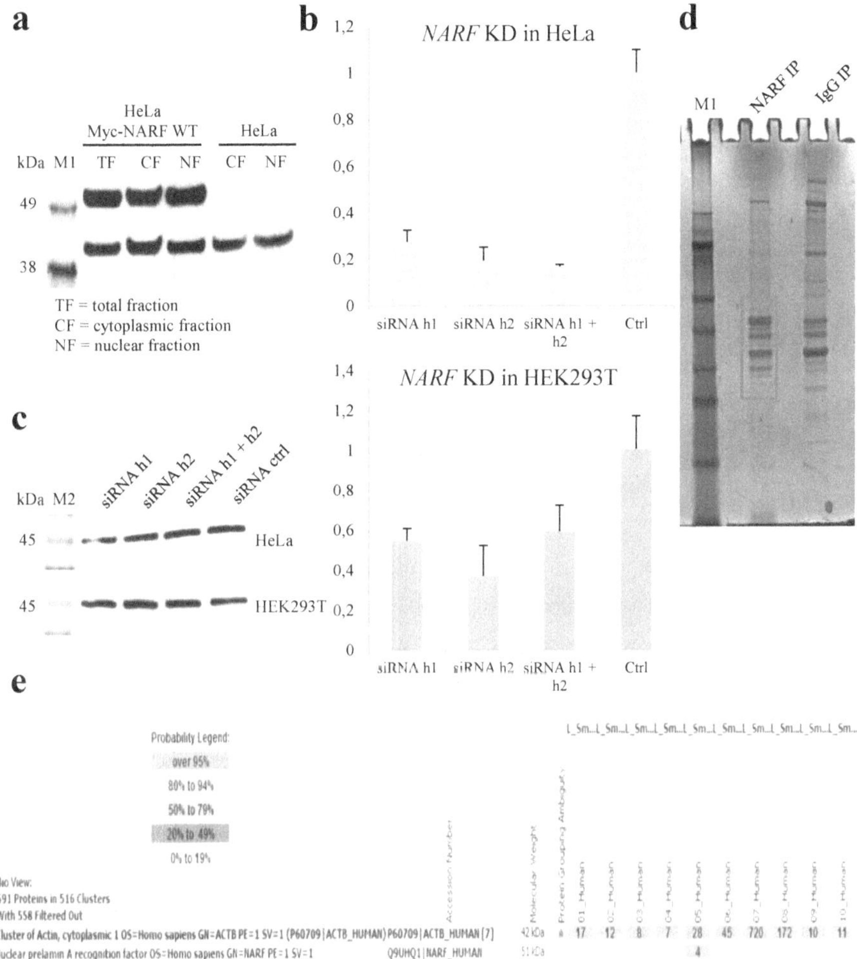

Figure 6: Generation of antibodies against mouse/human NARF protein. (a) Western blot analysis showing the specificity of the generated llama anti-NARF antibodies. The tested antibody recognised the exogenous overexpressed Myc-tagged NARF (upper band ~52 kDa) and a smaller endogenous protein (lower band ~42 kDa) in protein lysate extracted from *NARF*-overexpressing and WT HeLa cells. **(b)** qRT-PCR results demonstrating the efficiency of siRNA knock-down (KD) of *NARF* in HeLa and HEK293T cells. siRNA transfection reduced the expression of *NARF* to ~10-20% and ~40-60% in the HeLa and HEK293T cell lines, respectively. The values and associated error bars represent mean ± *SD* (*n*=3). **(c)** WB results indicating that

siRNA KD exerts no effect on the expression of the endogenous protein detected by the tested antibody. Cells treated with siRNA against NARF and the cells treated with scrambled siRNA were comparable in terms of the intensity of the signal. **(d)** Results of the immunoprecipitation (IP) assay performed with either the tested rabbit anti-NARF (left column) or anti-rabbit IgG (right column) antibodies on the protein lysate from the HeLa cells. The red box represents the gel fragments used for proteomics analysis. **(e)** The list of the proteins detected in the rabbit anti-NARF IP. The most abundant protein detected by the rabbit anti-NARF antibody was β-actin (1030 hits). NARF itself was also detected with the IP fraction, albeit with extremely low efficiency (4 hits). M1 = SeeBlue Plus2 Pre-Stained Protein Standard, M2 = Precision Plus Protein™ All Blue Pre-Stained Protein Standards.

5.3 Effect of the p.His367Arg mutation on the cellular localisation of NARF

The nuclear localisation of the NARF protein has been previously described (Barton and Worman, 1999). To validate this and to analyse the impact of the p.His367Arg mutation on protein localisation, I generated expression plasmids for wild-type (further referred as NARF WT) and mutant (further referred as NARF$^{p.H367R}$) NARF, both N-terminally tagged with a Myc-tag. I transfected HeLa cells with these expression constructs to overexpress WT or mutant NARF. Subsequently, I subjected the transfected cells to immunostaining with anti-Myc antibodies, and I analysed the subcellular localisation of NARF using fluorescence microscopy. Expression of Myc-tagged NARF WT in the HeLa cells resulted exclusively in a nuclear localisation, mainly co-localising with the nuclear envelop. By contrast, Myc-tagged NARF$^{p.H367R}$ was primarily present in the cytoplasm of the transfected cells, and it was no longer translocated to the nucleus as observed for NARF WT. In addition, NARF$^{p.H367R}$ was not evenly distributed in the cytoplasm, but aggregated within the cytoplasm, forming puncta and clump-like structures (Figure 7a). Subsequently, I confirmed these results independently through WB analyses. I subjected the proteins extracted from the HeLa cells overexpressing either NARF WT or NARF$^{p.H367R}$ to subcellular fractionation, and I isolated the cytoplasmic and nuclear fractions of the proteins and subjected them to WB analysis using anti-Myc-tag antibodies. I found that NARF WT was present in both the cytoplasmic and the nuclear fractions extracted from the HeLa cells, but I only detected NARF$^{p.H367R}$ in the cytoplasmic fraction of the extracted proteins (Figure 7b); this confirmed the results obtained by immunostainings.

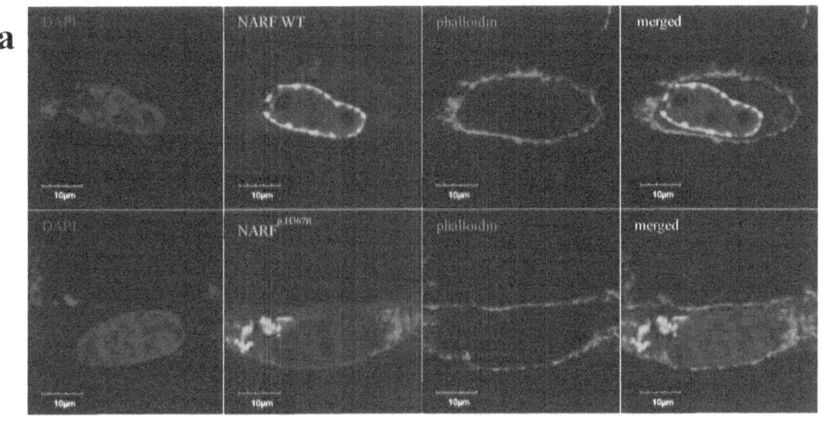

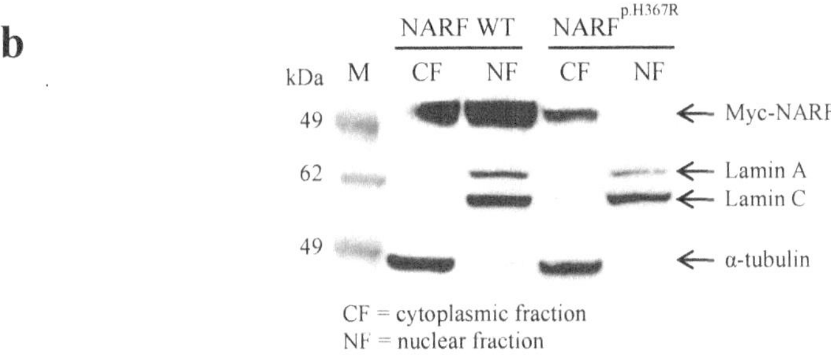

Figure 7: Cellular localisation of overexpressed WT NARF and mutant NARF^{p.H367R} proteins. (a) Immunofluorescence staining with the anti-Myc-tag antibodies (green) revealed expression of Myc-tagged WT NARF in the nucleus and expression of mutant Myc-tagged NARF^{p.H367R} in the cytoplasm. I used DAPI (blue) and phalloidin (red) staining to stain the DNA and to show the shapes of the cells, respectively. NARF WT was localised in the nucleus, mostly on the nuclear envelope, while mutant NARF^{p.H367R} was expressed in the cytoplasm and accumulated in aggregates. Scale bars = 10 µm. **(b)** Western blot analysis depicting the cellular distribution of the overexpressed WT NARF and mutant NARF^{p.H367R} proteins. The NARF WT protein was present in both fractions, but was predominantly expressed in the nuclear fraction, whereas mutant NARF^{p.H367R} was expressed exclusively in the cytoplasmic fraction. I used the anti-LMNA/C antibody and anti-α-tubulin as nuclear and cytoplasmic fraction controls, respectively. M = SeeBlue Plus2 Pre-Stained Protein Standard, CF = cytoplasmic fraction, NF = nuclear fraction.

Interestingly, the expression of NARF^{p.H367R} was always significantly lower than the expression of NARF WT. To determine whether the observed differences were attributable to inefficient transfection or the instability and degradation of the mutant protein, I transfected

HeLa cells with both forms of the NARF protein tagged with Myc-tag. I then extracted the cytosolic and nuclear protein fractions from 24, 48, and 72 hours after transfection and analysed them through WB using anti-Myc-tag antibodies. The WB analysis indicated that NARF WT was equally expressed over time, whereas the expression of mutant NARF dramatically decreased 48 hours after transfection. At 72 hours after transfection, NARF$^{p.H367R}$ could not be detected in either the lysates (Figure 8).

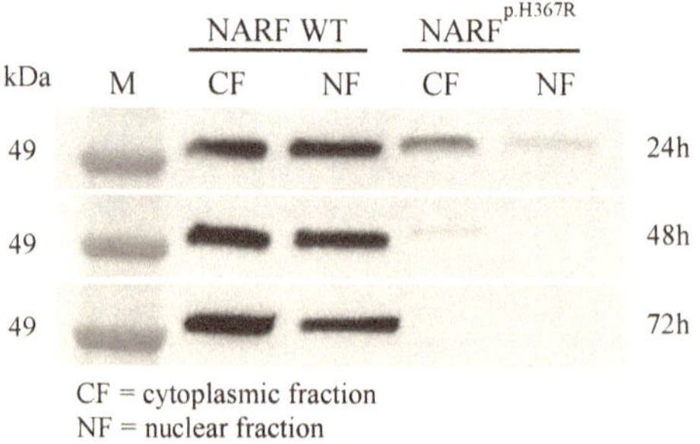

Figure 8: Stability of WT NARF and mutant NARF$^{p.H367R}$. Western blot results illustrating the expression levels of overexpressed WT and mutant NARF 24, 48, and 72 hours after the transfection of HeLa cells with pCMV-Myc-N plasmids expressing either NARF WT or NARF$^{p.H367R}$. I detected overexpressed proteins using anti-Myc-tag antibodies. The NARF WT protein was expressed 72 hours after transfection, while mutant NARF$^{p.H367R}$ was almost undetectable after 48 hours. M = SeeBlue Plus2 Pre-Stained Protein Standard, CF = cytoplasmic fraction, NF = nuclear fraction.

The severe effect induced by a single amino acid change in the protein sequence suggested that the conserved histidine at position 367 in NARF plays an important role for protein function. To further evaluate these findings, I designed a series of mutants, introducing a point mutation into the triplet encoding the histidine at position 367; I established six different mutant forms of NARF (named mutations 1-6, as enumerated in Table 31). Mutation 6 represented a negative control for this experiment. This represents a synonymous variant, c.1101C>T, not altering the amino acid (p.H367H).

Table 31: Amino acid changes in the NARF mutants. List of the mutations introduced into the NARF protein at amino acid position 367. The table reports the changes at the cDNA and protein levels and the characteristics of the newly introduced amino acids.

Name	Mutation	Amino Acid Substitution	Type of Amino Acid Change
Mutation 0 – patient's mutation	c.1100A>G	p.H367R	basic > basic aromatic > aliphatic
Mutation 1	c. 1100 A>C	p.H367P	basic > neutral aromatic > aliphatic
Mutation 2	c.1100 A>T	p.H367L	basic > neutral aromatic > aliphatic
Mutation 3	c.1099 C>G	p.H367D	basic > acidic aromatic > aliphatic
Mutation 4	c.1101 C>G	p.H367Q	basic > neutral aromatic > aliphatic
Mutation 5	c.1099 C>T	p.H367Y	basic > neutral aromatic > aromatic
Mutation 6 – negative control	c.1101 C>T	p.H367H	–

I cloned all the mutants' open reading frames (ORFs; 1-6) into the pCMV-Myc-N vector. Next, I transfected HeLa cells with all the constructs and subjected them to immunostaining with anti-Myc-tag antibodies. Using fluorescence microscopy, I examined the subcellular localisation of all six of the mutants. Like NARF[p.H367R], all the additional mutants (NARF[p.H367P], NARF[p.H367L], NARF[p.H367D], NARF[p.H367Q], and NARF[p.H367Y]) accumulated within aggregates in the cytoplasm of the transfected cells (Figure 9). Meanwhile, the negative control represented by a synonymous change (mutation 6, Table 1) was localised in the nucleus and associated with the nuclear envelope, as observed for NARF WT (Figure 9). These results substantiated the important role of the histidine at position 367 in the cellular localisation of NARF. All changes at position 367 resulted in impaired nuclear import.

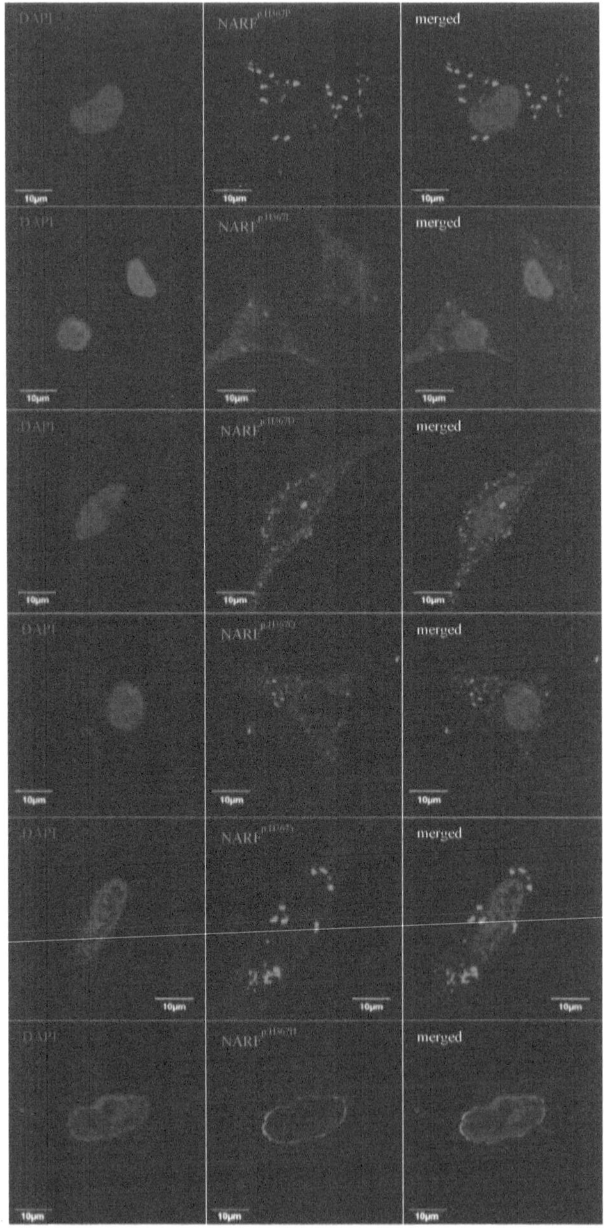

Figure 9: Subcellular localisation of NARF mutants carrying different amino acids at position p.H367 (NARF^{p.H367P}, NARF^{p.H367L}, NARF^{p.H367D}, NARF^{p.H367Q}, and NARF^{p.H367Y}). I employed immunofluorescence staining with anti-Myc-tag antibodies (red) to detect the Myc-tagged mutant proteins in the HeLa cells. All the

NARF mutants were localised in the cytoplasm, while NARF$^{p.H367H}$, which served as the control, was localised in the nucleus. I counterstained the nuclei with DAPI (blue). Scale bars = 10 µm.

5.4 Identification of novel NARF interaction partners

To expand our knowledge regarding the potential function of NARF, I attempted to identify novel NARF interaction partners. In collaboration with the Hybrigenics Services, a yeast-two-hybrid (Y2H) assay was performed. We used full-length NARF as bait and screened interactions using human ventricle and embryo heart cDNA libraries. In this assay, we were not able to verify interaction with pre-lamin A, which has been previously reported (Barton and Worman, 1999). Still, we identified two novel protein interaction partners with very high confidence, that is, NARF and CBX5 (HP1-α; Figure 10). The presence of NARF itself suggested the formation of homodimers by NARF. The second interaction partner, CBX5, is a mammalian heterochromatin protein that interacts with many nuclear proteins, including lamins. CBX5 also interacts with lamin A, and it has been associated with premature ageing in a mouse model (Liu et al., 2014). I used both putative interaction partners for further analysis.

Clone Name	Type Seq	Gene Name (Best Match)	Start..Stop (nt)	Frame	Sense	%Id 5p	%Id 3p	PBS
pB27_A-11	5p/3p	Homo sapiens - ADGRE2	642..1524	IF		99.5	71.4	
pB27_A-12	5p/3p	Homo sapiens - ADGRE2	642..1524	IF		99.8	79.4	
pB66_A-20	5p/3p	Homo sapiens - ATAD3C	2920..2662	??	N	90.1	90.1	N/A
pB66_A-1	5p/3p	Homo sapiens - ATP7B	363..1127	IF		99.6	99.7	
pB27_A-15	5p/3p	Homo sapiens - ATRX	4506..4790	IF		100.0	100.0	
pB66_A-27	5p	Homo sapiens - CBX5	87	IF		96.9		A
pB66_A-16	5p	Homo sapiens - CBX5	87	IF		99.7		A
pB66_A-13	5p/3p	Homo sapiens - CBX5	246..1081	IF		99.1	96.0	A
pB27_A-18	5p/3p	Homo sapiens - CBX5	246..1081	IF		99.8	99.5	A
pB27_A-7	5p/3p	Homo sapiens - CBX5	297..1549	IF		100.0	99.8	A
pB66_A-32	5p/3p	Homo sapiens - CDC73	2352..419	??	N	99.6	99.5	N/A
pB27_A-5	5p/3p	Homo sapiens - CREBZF	498..999	IF		100.0	99.6	
pB66_A-3	5p/3p	Homo sapiens - ECHDC2	14..-770	??	N	98.6	100.0	N/A
pB66_A-12	5p/3p	Homo sapiens - HERC2	10176..10021	??	N	100.0	100.0	N/A
pB27_A-8	5p/3p	Homo sapiens - HEY2	60..897	IF		99.4	99.8	
pB66_A-8	5p/3p	Homo sapiens - HIPK2 variant 1	2433..2784	IF		100.0	100.0	
pB66_A-6	5p/3p	Homo sapiens - HIPK2 variant 1	2433..2784	IF		100.0	99.4	
pB66_A-10	5p/3p	Homo sapiens - HIPK2 variant 1	2433..2784	IF		100.0	100.0	
pB66_A-15	5p/3p	Homo sapiens - HIPK3	2187..2739	IF		99.3	92.0	
pB66_A-17	5p/3p	Homo sapiens - MYH7	5461..5048	??	N	100.0	100.0	N/A
pB27_A-14	5p/3p	Homo sapiens - NARF	1113..1366	IF		96.2	75.0	A
pB27_A-16	5p/3p	Homo sapiens - NARF	1155..1464	IF		100.0	100.0	A
pB27_A-13	5p/3p	Homo sapiens - NARF	1155..1464	IF		100.0	100.0	A
pB27_A-2	5p/3p	Homo sapiens - NARF	1155..1464	IF		100.0	100.0	A
pB27_A-10	5p/3p	Homo sapiens - NARF	1155..1464	IF		100.0	100.0	A
pB66_A-4	5p/3p	Homo sapiens - ORC4	4781..5218	OOF2		90.9	90.9	N/A
pB66_A-19	5p	Homo sapiens - PALLD	5054..5673	OOF2		85.6		N/A
pB66_A-25	5p/3p	Homo sapiens - PIAS1	1260..1527	IF		100.0	100.0	

Figure 10: Identification of novel NARF interaction partners using a yeast-two-hybrid (Y2H) screening.
The summary of the Y2H screening results provided by the Hybrigenics Services. The Y2H screening identified two potential interaction partners of NARF: CBX5 and NARF itself (red frames). Based on the global Predicted Biological Score (PBS), these interactions were predicted with very high confidence (A in red square). Clone names: pB = bait vector; A = prey clone number; Type Seq indicates whether 5p and/or 3p sequences are available for prey identification; Start/Stop = position of the 5p and 3p prey fragment ends relative to the position of the ATG start codon (A = 0); IF = in frame with the Gal4 Activation Domain; OOF1, OOF2 = out of frame; N = antisense orientation with respect to the reference sequence; %Id 5p/3p indicates the % identity of the prey fragment sequences with the gene reference sequence; PBS = Predicted Biological Score, which indicates the confidence in an interaction; A = very high confidence in the interaction; D = moderate confidence in the interaction; E = interactions involving highly connected prey domains, non-specific interactions; N/A = not applicable.

To validate the co-localisations of likely interacting proteins, I transfected HeLa cells with hEF1α-GFP NARF WT plasmids. Subsequently, I subjected the cells to immunostaining using anti-LMNA/C or anti-CBX5 antibodies to define the cellular localisations of endogenous lamin A and CBX5, respectively. Using fluorescence microscopy, I confirmed the cellular co-localisations of all three proteins. All interaction partners were present in the nuclear compartment of the transfected cells. Additionally, NARF and lamin A co-localised within the nuclear envelope (Figure 11). WB analyses further substantiated these results. I subjected protein extracts from the HeLa cells transfected with either Myc-tagged NARF WT or Myc-tagged NARF$^{p.H367R}$ to WB analysis and probed the membranes with anti-Myc-tag, anti-lamin A/C, and anti-CBX5 antibodies. The results confirmed that NARF WT, but not NARF$^{p.H367R}$, co-localises in nuclear protein fractions. I used proteins extracted from non-transfected HeLa cells as a control; they demonstrated expression of endogenous lamin A and CBX5 proteins, but not exogenous NARF-Myc proteins. The presence of mutant NARF$^{p.H367R}$ exerts no impact on the expression or subcellular distribution of either lamin A or CBX5 (Figure 12). Both proteins remained present in the nuclear fractions of the cell lysates.

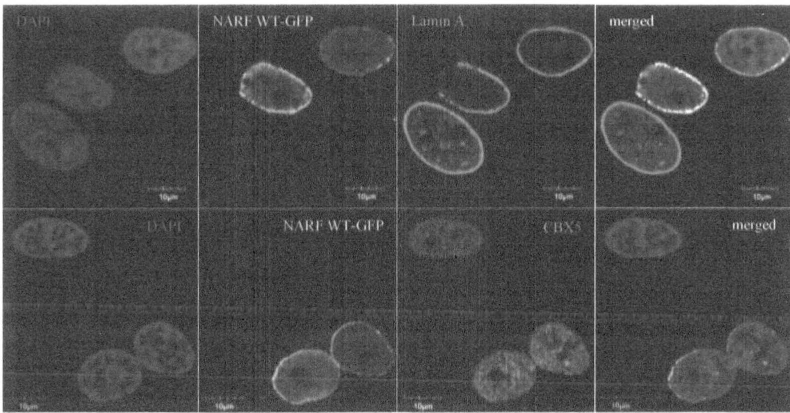

Figure 11: Co-localisations of NARF and its interaction partners lamin A and CBX5. Immunofluorescent images of HeLa cells transfected with WT NARF-GFP (green). I employed anti-LMNA/C and anti-CBX5 antibodies to analyse the localisations of endogenous lamin A and CBX5, respectively (red). NARF co-localises with both the tested interaction partners within the nucleus. I counterstained the nuclei with DAPI (blue). Scale bars = 10 µm.

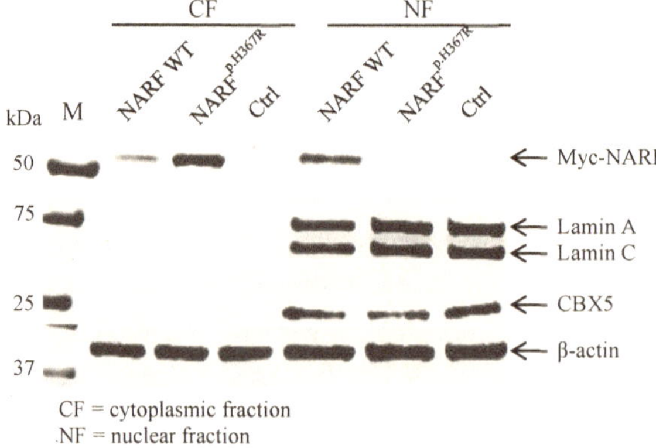

Figure 12: Cellular distribution of NARF WT, mutant NARF$^{p.H367R}$, lamin A/C, and CBX5. Western blot results depicting the cellular distribution of overexpressed NARF WT, mutant NARF$^{p.H367R}$, and endogenous lamin A and CBX5 in cytosolic and nuclear protein fractions from HeLa cells. The NARF WT protein was present in both fractions but was predominantly expressed in the nuclear fraction, while mutant NARF$^{p.H367R}$ was expressed exclusively in the cytoplasmic fraction. Lamin A and CBX5 were present only in the nuclear fraction. Overexpression of mutant NARF$^{p.H367R}$ altered neither the expression nor the cellular distribution of lamin A and CBX5. Protein extracts from non-transfected cells (Ctrl) served as a control. I used the β-actin antibody as a control for protein loading. M = Precision Plus Protein™ All Blue Pre-Stained Protein Standards.

To examine the direct interactions between NARF and lamin A, I conducted co-immunoprecipitation (CoIP) experiments. I transfected HeLa cells with expression constructs for Myc-tagged NARF WT; I then extracted proteins and used the nuclear fractions in the CoIP experiments. Using magnetic beads conjugated with anti-Myc-tag antibodies and the nuclear fractions, I initiated immunoprecipitation (IP) of the proteins. After allowing the IP to proceed overnight, I eluted the proteins from the beads, denaturised them, and subjected them to WB analysis using anti-LMNA/C antibodies. Since lamin A was co-eluted from the protein extracts alongside with Myc-NARF, the results evinced a direct interaction between NARF and lamin A. To exclude unspecific binding, I performed the same procedure using beads conjugated with mouse IgG as a negative control (Figure 13a). I executed the same CoIP procedures to study the direct interactions between NARF and CBX5. Nevertheless, my efforts to co-precipitate NARF with CBX5 were unsuccessful (data not shown); I therefore decided to switch the methods and to use a pull-down assay to investigate the direct interactions between NARF and CBX5. For this purpose, I used cell lysates from HeLa cells overexpressing Myc-tagged NARF proteins; additionally, I used recombinant His-tagged CBX5. Recombinant CBX5-His was expressed in the bacterial BL21 strain, and the presence of the His-tag enabled purification of both, the CBX5 protein from the bacterial lysates, and subsequent purification of the protein complexes. I extracted the nuclear and cytoplasmic fractions of the HeLa cells overexpressing NARF WT or NARF$^{p.H367R}$ proteins, and I performed a pull-down assay overnight, using the purified recombinant CBX5-His protein and the nuclear and cytoplasmic lysates. Using HisPur™ resins, I purified the protein complexes again; I then subjected the proteins to WB analysis using anti-Myc-tag antibodies. Since NARF was pulled down alongside with CBX5-His, the results confirmed a direct interaction. Moreover, this experiment demonstrated that the mutations do not disrupt the interaction between NARF and CBX5, as NARF$^{p.H367R}$ retained the ability to bind and be pulled down together with CBX5-His (Figure 13b).

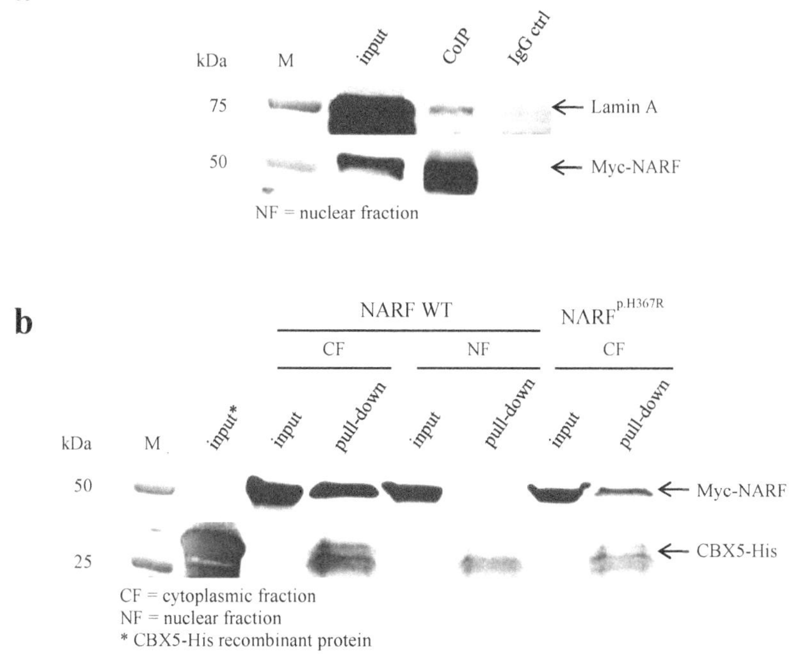

Figure 13: NARF interaction partners. (a) Western blot results of the co-immunoprecipitation experiments conducted in HeLa cells overexpressing NARF WT. Using the anti-Myc-tag antibody, I performed immunoprecipitation of the NARF-lamin A complex in the nuclear fractions; subsequently, I conducted WB analysis using anti-lamin A/C antibodies. Lamin A was co-precipitated alongside with NARF WT. As a negative

control, I performed IP with the IgG antibody. **(b)** Western blot results of the pull-down experiments conducted with the recombinant CBX5-His protein and protein extracts from HeLa cells overexpressing NARF WT and NARF[p.H367R]. I incubated the protein extracts with recombinant CBX5-His; subsequently, the protein complexes were purified through resins binding the His-tag. I conducted WB analysis of the purified proteins, using anti-Myc tag antibodies. Both the NARF and NARF[p.H367R] proteins were pulled down alongside with the recombinant CBX5 protein. CF = cytoplasmic fraction, NF = nuclear fraction, M = Precision Plus Protein™ All Blue Pre-Stained Protein Standards.

To verify the direct interactions between NARF and lamin A as well as CBX5, I designed and prepared new expression plasmids for a bimolecular fluorescence complementation (BiFC) assay. Briefly, this experiment requires co-transfection of cells with plasmids that express potential interaction partners fused to the C- or N-terminal fragment of Venus fluorescent protein. Fragments of reporter protein are linked to the C- or N-terminus of the examined proteins with flexible linkers that enable reformation of the native structure of the reporter protein and emission of a fluorescent signal in the case of direct interactions between the tested proteins (Kerppola, 2006). I co-transfected HeLa cells with plasmids expressing NARF WT fused to the N-terminal fragment of the reporter Venus protein (VN) and NARF WT, lamin A, or CBX5 fused to the C-terminal fragment of the reporter Venus protein (VC), respectively. Using fluorescence microscopy, I observed direct NARF-NARF, NARF-lamin A, and NARF-CBX5 interactions, which resulted in green fluorescent signals emitted by the reconstituted Venus protein (VN + VC) (Figure 14). Taken together, I was able to identify and confirm CBX5 and NARF itself as novel interaction partners.

NARF WT (VC)   LMNA (VC)   CBX5 (VC)

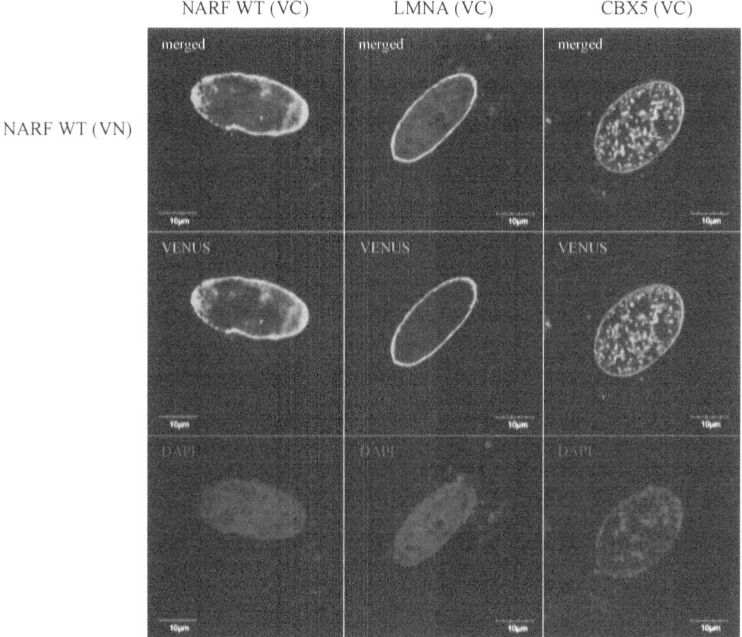

Figure 14: Bimolecular fluorescence complementation assay of NARF and its interaction partners. Immunofluorescent photographs illustrating the results of the bimolecular fluorescence complementation (BiFC) assay testing the direct interactions between NARF WT and its potential interaction partners (NARF, lamin A, and CBX5). Interactions between the proteins enabled reconstitution of the Venus protein (VN + VC) and emission of fluorescent signals (green). I counterstained the nuclei with DAPI (blue). Scale bars = 10 µm.

5.5 Dominant negative effect of NARF$^{p.H367R}$

The impaired cellular localisation of NARF$^{p.H367R}$ and the homodimerisation of NARF raised the questions of a possible dominant negative effect of NARF$^{p.H367R}$ on NARF WT. To investigate this, I co-transfected HeLa cells with both WT and mutant forms of the NARF protein tagged with different tags (Myc or GFP) to distinguish between overexpressed proteins. Co-expression of Myc- and GFP-tagged WT proteins resulted in fluorescence signals in the nucleus, primarily within the nuclear envelope (Figure 15a). By contrast, co-expression of WT and mutant proteins still co-localised, but this occurred predominantly outside the nucleus (Figure 15b). Similarly, co-expression of GFP- and Myc-tagged mutant proteins also indicated co-localisation exclusively within the cytoplasm of transfected cells (Figure 15c).

To confirm the dominant negative effect of NARF$^{p.H367R}$, I performed a bimolecular fluorescence complementation (BiFC) assay. I co-transfected HeLa cells with plasmids expressing NARF WT fused to the C-terminal fragment of the reporter Venus protein (VC) and the NARF$^{p.H367R}$ protein fused to the N-terminal fragment of the reporter Venus protein (VN). As depicted in Figure 15d, mutant NARF$^{p.H367R}$ interacts with WT NARF, but the localisation of these NARF complexes is mainly restricted to the cytoplasm (Figure 15d).

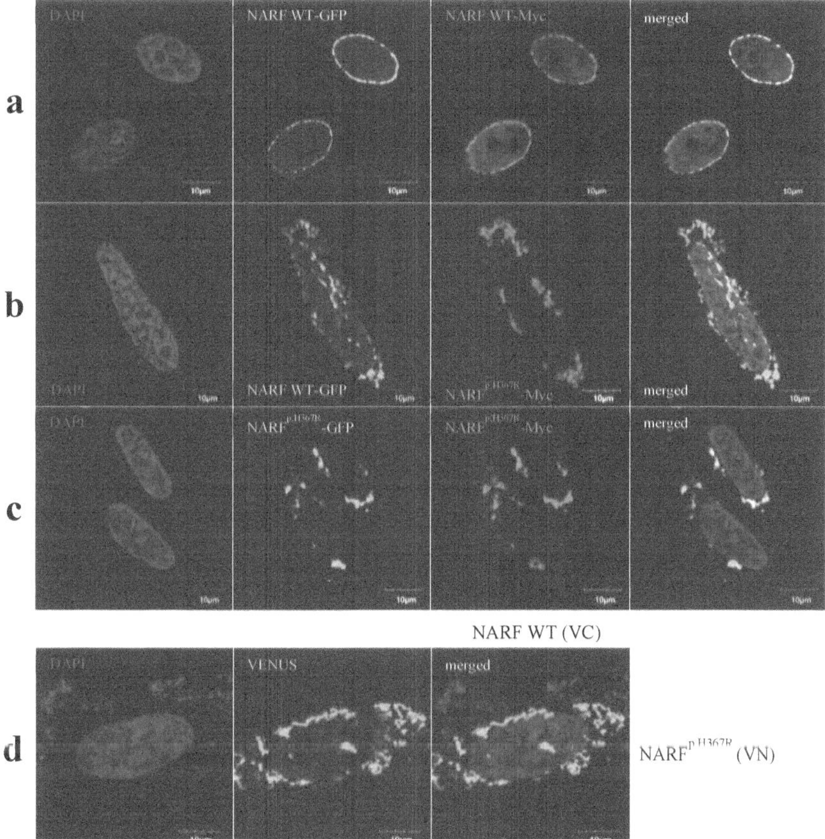

Figure 15: A dominant negative effect of the mutant NARF$^{p.H367R}$ protein. **(a)** Immunofluorescent analysis of the HeLa cells co-transfected with GFP- and Myc-tagged NARF WT (green and red, respectively). The GFP- and Myc-tagged NARF WT co-localised at the nuclear envelope. **(b)** Immunofluorescent analysis of the HeLa cells co-transfected with GFP-tagged NARF WT (green) and Myc-tagged NARF$^{p.H367R}$ (red). WT and mutant NARF co-localised but were detected primarily in the cytoplasm and only to a minor extent in the nucleus. **(c)** Immunofluorescent analysis of HeLa cells co-transfected with GFP- and Myc-tagged NARF$^{p.H367R}$ (green and red, respectively). The mutant proteins co-localised exclusively outside the nucleus. **(d)** BiFC assay verifying the dominant negative effect of the p.His367Arg mutation. The NARF$^{p.H367R}$ was able to interact with the NARF WT, thereby limiting the protein complex localisation to the cytoplasm. I counterstained the nuclei with DAPI (blue). Scale bars = 10 μm.

I demonstrated that the mutation does not disrupt the NARF-NARF interaction; however, NARF$^{p.H367R}$ exerts an effect on NARF WT, as it alters its subcellular localisation. The obtained results might suggest that the conserved histidine at position 367 possibly plays a role in protein complex trafficking and proper localisation.

5.6 Is NARF function evolutionarily conserved?

NARF is a highly conserved protein. To analyse whether human *NARF*, like its yeast homologue *NAR1*, boasts an evolutionarily conserved function, I performed a complementation assay in yeast. To this end, I employed a decreased abundance by mRNA perturbation (DAmP) Nar1 yeast strain that allows for generating hypomorphic alleles of essential yeast genes. Briefly, the 3' untranslated region (3'UTR) of a gene is disrupted through insertion of an antibiotic resistance cassette, which engenders the destabilisation of the transcript and the reduction of mRNA levels (Breslow et al., 2008). Because there are two Nar1 homologues—that is, NARF and NARFL—in mammals, I implemented cDNA encoding for both proteins and tested their ability to rescue the Nar1-deficient phenotype in yeast. In a first approach, I cloned yeast codon-optimised ORFs of *NARF*, *NARFL*, and *NAR1* into the p415 mTag BFP2 vector. To test the expression of the human proteins in yeast, I transformed the WT By4741 yeast strain with p415 mTag BFP2 -NARF, -NARFL, and -Nar1 plasmids. The transformation of WT yeast resulted in detectable, but low expression of all three recombinant proteins in the logarithmic phase of yeast growth. In the stationary phase of growth, however, NARF and NARFL aggregated in foci (Figure 16), whereas Nar1 was evenly distributed in the cytoplasm.

By4741
+ p415 mTag BFP2-X

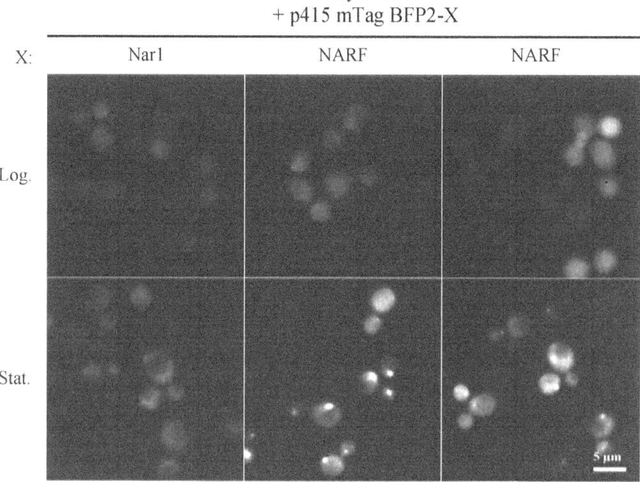

Figure 16: Expression of BFP-tagged Nar1, NARF, and NARFL proteins in the WT By4741 yeast strain. I transformed yeast strain By4741 with p415 mTag BFP2 plasmids containing yeast *NAR1* and yeast codon-optimised ORFs of human *NARF* and *NARFL* genes. Using fluorescence microscopy, I detected the expressed proteins, which were tagged with blue fluorescent protein (BFP2). All three recombinant proteins were equally distributed within the cells during the logarithmic phase of yeast growth. In the stationary phase, the human NARF and NARFL proteins aggregated in foci. Scale bar = 5 µm.

In a next step, I tested the different culture conditions to find conditions that exert an effect on the DAmP Nar1 strain, but are neutral for growing WT yeast. I cultured WT and DAmP Nar1 yeast strains either at different temperatures or on selective plates supplemented with either hydrogen peroxide (H_2O_2) or copper sulphate ($CuSO_4$). Preliminary experiments have indicated that the presence of hydrogen peroxide or copper sulphate, but not a higher temperature alone, is able to prevent the DamP Nar1 strain from growing (Figure 17a). I then performed a complementation assay on the plates supplemented with hydrogen peroxide. Still, none of the tested proteins were able to rescue the DAmP Nar1 phenotype (Figure 17b). Neither transformation with yeast Nar1, nor transformation with NARF or NARLF exerted any effect on the growth of this DAmP Nar1 strain. This result was observed independent of the presence of hydrogen peroxide. The negative results obtained in the experiment with the positive control plasmids (p415 mTag BFP2 Nar1) and the observed aggregation of BFP2-fused human proteins suggest that BFP may jeopardise the proper folding and function of tagged proteins. To rule out a detrimental impact of BFP2 on the expression of the Nar1, NARF, and NARFL proteins, I repeated the complementation assay using the DAmP Nar1 strain and untagged Nar1, NARF, and NARFL. Again, none of the proteins proved able to rescue the growth of DAmP Nar1 yeast (data not shown).

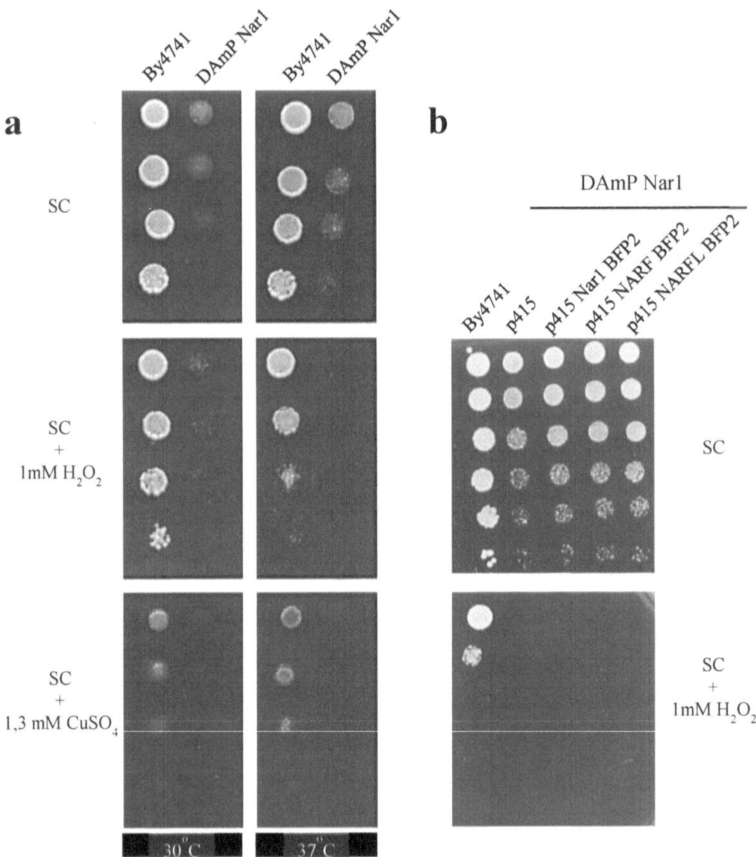

Figure 17: Complementation assay performed in the DAmP Nar1 yeast strain. (a) Identification of conditions that impair the growth of the DAmP Nar1 yeast strain. The growth of DAmP Nar1 is suppressed by H_2O_2 or $CuSO_4$ but not by increased temperature. **(b)** The growth of DAmP Nar1 can be rescued neither by the yeast Nar1 protein nor by the human homologues, NARF and NARFL. SC = Synthetic complete medium.

In light of these results, I concluded that the DAmP Nar1 strain and the selection conditions are not optimal. Therefore, I changed the yeast strains and used the ΔNar1 strain, in which the *NAR1* gene has been deleted (KO). Because *nar1*-KO is lethal, I transformed the ΔNar1 yeast with the p416 Nar1 plasmid. Expression of Nar1 from this plasmid is driven by the MET25 promoter. In media lacking methionine, the MET25 promoter induces high expression of Nar1, thus enabling normal growth of ΔNar1 yeast (Mumberg et al., 1994). Additionally, the p416 plasmid encodes for the *URA3* gene. URA3 is an orotidine-5'-phosphate decarboxylase (ODCase), which is an enzyme that catalyses one step in the synthesis of pyrimidine ribonucleotides (Flynn and Reece, 1999). ODCase can also convert 5-fluororotic acid (5-FOA) into the toxic compound 5-fluorouracil (Boeke et al., 1984). This allows for selecting against yeast carrying the *URA3* gene. In the presence of 5-FOA, these cells will die. I applied this model to perform a further complementation assay. I transformed ΔNar1-p416 Nar1 yeast with p415 plasmids encoding for either Nar1 or its human homologues, NARF and NARFL. I then cultured the transformed yeast on 5-FOA selective plates, thereby eliminating cells that express Nar1 from the p416 plasmid. This enabled me to test the effect of the transfected factors encoded within the p415 plasmid (Nar1, NARF, and NARFL). In the first trial, I transformed ΔNar1-p416 Nar1 yeast with p415 encoding for untagged or BFP2-tagged Nar1, NARF, or NARFL. In these cases, the Nar1 protein proved able to rescue the growth of the ΔNar1 yeast, whereas the yeast transformed with either NARF or NARFL died on undergoing the 5-FOA treatment (Figure 18a). In addition, to help stabilise the human proteins and possibly improve their folding and/or solubility, I tested another tag: tandem A protein (ZZ-tag). I obtained similar results: I was only able to rescue the growth of ΔNar1 with the Nar1 protein (Figure 18b, upper panel). Since the proteins expressed from the p415 were under the control of the MET25 promoter, which is sensitive to methionine concentrations, I decreased the concentration of methionine in the culture medium from 20 mg/L to 5 mg/L. Nevertheless, lowering the methionine concentration exerted no effect on the inability of either human NARF or human NARFL to rescue ΔNar1 yeast (Figure 18b, lower panel). Taken together, these results indicate that neither NARF nor NARFL is able to rescue ΔNar1 yeast; thus, these results suggest that yeast Nar1 fulfils a different functional role than the one fulfilled by human NARF and NARFL.

a

By4741 *nar1*Δ::NAT p416 Nar1 +

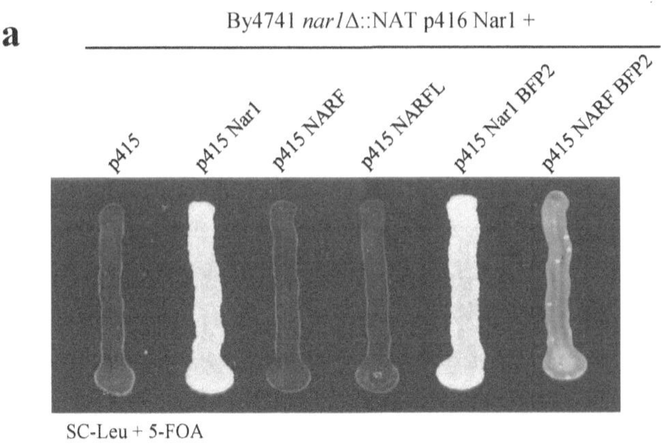

b

By4741 *nar1*Δ::NAT p416 Nar1
+ p415 2x ZZ-tag TEV-X

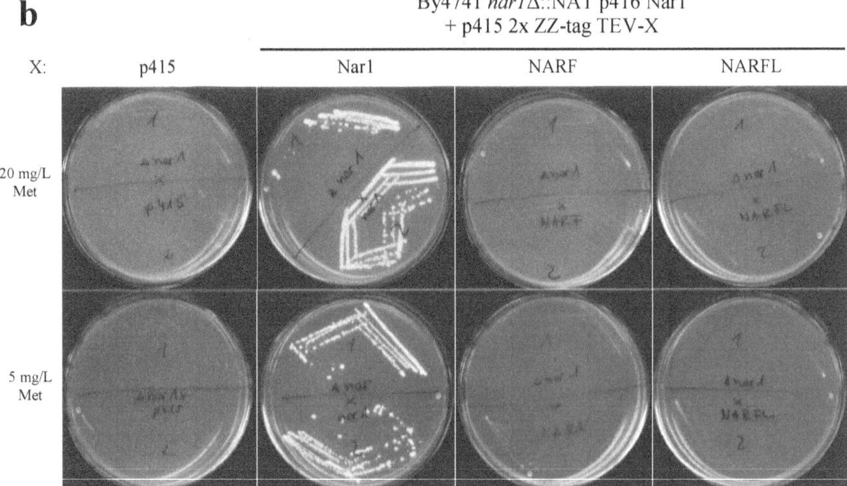

Figure 18: Complementation assay performed in *nar1*-KO yeast (ΔNar1). (a) Transformation of *nar1*-KO yeast with untagged or BFP2-tagged Nar1, NARF, or NARFL. NARF and NARFL both proved unable to recover the yeast growth. **(b)** I obtained the same results when I used ZZ-tagged proteins in the presence of normal and decreased methionine concentrations. Transformation with yeast Nar1 served as the control. In both experiments, the yeast Nar1 protein rescued the ΔNar1 phenotype. SC = Synthetic complete medium.

5.7 Generation of a Narf$^{p.H373R}$ knock-in mouse as a model for the progeroid syndrome

To determine the effect of the identified p.His367Arg mutation in *NARF* and to analyse NARF function in general, I aimed to generate a CRISPR/Cas9-mediated knock-in (KI) mouse model. The patient's mutation corresponds to position 1118 (c.1118A>G, p.His373) in exon 10 of *Narf*, and it converts the histidine at position 373 to arginine (p.His373Arg). To introduce this particular mutation into *Narf*, I used two CRISPR vectors (CRISPR Narf #2 and CRISPR Narf #3) encoding the essential components for the CRISPR/Cas9 approach, such as the guide RNA (gRNA) complement to the target region in exon 10 adjacent to the amino acid p.His373, the Cas9 nuclease that provides DNA cutting and formation of double-strand breaks (DSBs) within the target site, and red fluorescent protein (RFP) for easier selection of transfected cells. CRISPR Narf #2 and CRISPR Narf #3 plasmids differ slightly in their gRNA sequences, but both target the region within exon 10 near the codon of amino acid p.His373. In a preliminary experiment, I validated the specificity and efficiency of the provided plasmids. Briefly, I transfected mouse embryonic stem (mES) cell line EDJ #22 with CRISPR Narf #2 or CRISPR Narf #3 plasmids. After implementing fluorescence-activated cell sorting (FACS), I cultured RFP-positive cells to obtain mES cell colonies. I cultured each colony separately to give rise to an mES cell line. In this way, I established 11 and 33 mES cell lines after transfection with CRISPR Narf #2 and CRISPR Narf #3, respectively. I genotyped each mES cell line through Sanger sequencing. Most of the mES lines presented with small 'indels' or substitution mutations at the targeted position in exon 10 of *Narf* (Table 32), thus demonstrating that the tested CRISPR plasmids efficiently introduced DSBs and activated DNA repair mechanisms.

Table 31: List of mutations obtained after CRISPR/Cas9 transfection. Summary of the mutations introduced by repairing of DNA double-stranded breaks (DSBs) generated by CRISPR Narf #2 and CRISPR Narf #3. DNA changes, corresponding protein changes together with genetic status of each clone are listed in the table.

EDJ #22 CRISPR Narf #2			
Clone	cDNA status	Protein status	Genetic status
#1	WT	WT	WT
#2	WT	WT	WT
#3	c.1039 C>T c.1055 C>G c.1058 A>T c.1089 G>C	p.A353G p.Y354F p.Q363H p.L365I	compound heterozygote

Clone	cDNA status	Protein status	Genetic status
	c.1093 C>A c.1098 G>C c.1101 G>A c.1110-1120 del (TCCCATACCACT)	p.K366N fs*432	
	c.1114-1118 del (TACCA)	p.Y372L fs*434	
#5	c.1114-end of exon 10? del	p.372-379 del (YHFVEVLA) p.381-383 del (PRG)	heterozygote
	WT	WT	
#6	WT	WT	heterozygote
	c.1127 A>G	p.E376G	
#8	WT	WT	WT
#9	c.998 A>G c.1117 C>T c.1118-1119 del (AC)	p.D333G fs*435	compound heterozygote
	c.1117 C>T c.1118-1119 del (AC)	p.H373F fs*435	
#10	c.1109-1131 del c.1144 A>G	p.L370R fs*378	compound heterozygote
	c. 1117-1118 CA>TT c. 1119 del (C)	p.H373F fs*385	

EDJ #22 CRISPR Narf #3

Clone	cDNA status	Protein status	Genetic status
#1	WT	WT	WT
#3	WT	WT	WT
#4	c.1011 T>A c.1121-1122 del (TT)	p.F374C fs*434	compound heterozygote
	c.1024 A>G	p.N342D	
#5	WT	WT	WT
#7	WT	WT	WT
#8	c.1121-1133 del (TTGTGGAGGTGCT)	p.F374S fs*381	homozygote
#9	c.1103-1126 del (GCAAACTCCCATACCACTTGTGG)	p.368-375 del (GKLPYHFV)	compound heterozygote
	c.1123-1124 ins (TT)	p.V375L fs*386	
#12	c.1122-1126 del (TGTGG)	p.F374L fs*434	homozygote
#13	c.1122-1128 del (TGTGGAG)	p.F374L fs*383	homozygote
#14	c.1122-1128 del (TGTGGAG)	p.F374L fs*383	homozygote
#15	c.1122-1126 del (TGTGG)	p.F374L fs*434	homozygote
#17	c.1117-1121 del (CACTT)	p.H373C fs*434	homozygote
#18	c. 1105-1126 del c.1130 del (T)	p.K369R fs*428	homozygote
#19	c.1122-1124 del (TGT)	p.F374L p.375 del (V)	compound heterozygote
	c.1117-1121 del (CACTT)	p.H373C	

		fs*434	
#20	c.-35-1126 del		compound heterozygote
	c.1104-1126 del	p.K369G fs*428	
#22	c.1038 A>G c.1119-1124 del (CTTTGT)	p.H373Q p.374-375 del (FV)	compound heterozygote
	c. 1105-1126 del c.1130 del (T)	p.K369R fs*428	
#23	c.1092-1121 del	p.K364N p.365-374 del (LKKGKLPYHF)	homozygote
#25	c.1144 A>C c.1145 G>C c.1146-1147 ins (AG)	p.R382P fs*386	heterozygote
	WT	WT	
#26	WT	WT	WT
#27	c.1122 del (T)	p.F374L fs*385	homozygote
#29	c.1122 del (T)	p.F374L fs*385	compound heterozygote
	c.1091-1141 del	p.K364T p.365-381 del (LKKGKLPYHFVEVLACP)	
#31	c.1070-1120 del	p.357-373 del (NIQNMIQKLKKGKLPYH) p.F374I	compound heterozygote
	c.1108-1121 del	p.L370C fs*431	
#32	WT	WT	heterozygote
	c.1104-1147 del + 2 del (int 10-11)	p.K369L fs*421	
#33	c.1092-1121 del	p.K364N p.365-374 del (LKKGKLPYHF)	homozygote
#11	WT	WT	heterozygote
	c. 1112-1117 del (CATACC)	p.371-372 del (PY)	

Compound heterozygosity observed in some mES clones is attributable to different repair mechanisms occurring after DSBs. DSBs are predominantly repaired through homology-directed repair (HDR) and non-homologous end joining (NHEJ) mechanisms. The activation of different repair mechanisms on two alleles and/or combined DNA repair on one allele can introduce many small (often single) deletions, insertions, and substitutions, which in turn lead to the generation of compound heterozygote clones carrying different mutations on each allele at the target region.

In a next step, I tried to introduce the *Narf* c.1118A>G mutation into the mouse genome. For this purpose, I designed and prepared a special homologous recombination DNA (HRD) template, which was co-transfected with the CRISPR Narf#3 plasmid and served as a template for HDR. The HRD template encompassed the position of the desired mutation (*Narf* c.1118A>G) and 100 nucleotides flanking homologous arms. To protect the HRD template from Cas9 cutting, I introduced an additional silent mutation (*Narf* c.1116C>T) in a CRISPR Narf#3-specific PAM sequence. I co-transfected EDJ #22 mouse mES cells with CRISPR Narf #3 and the HRD template (Figure 19a, left path). After 24 hours, I collected RFP-positive cells through FACS. I then cultured them further to generate mES cell colonies and, finally, mES cell lines. In total, I was able to establish 284 mES cell lines. I genotyped each line via PCR and subsequent Sanger sequencing. Out of these 284 mES lines, I obtained 6 mES cell lines with a compound heterozygous (c.1118 A>G plus additional mutations as a result of combined HDR and NHEJ repair mechanisms) and 6 mES cell lines with a homozygous knock-in (KI hom; Figure 19a). Subsequently three KI hom clones were injected into blastocysts at Max Planck Institute of Experimental Medicine, Göttingen; this resulted in 19 chimeric pups. Most of them were low-grade chimeras, with chimerism varying between 5% and 25%, and only one male exhibited 60% chimerism (Figure 19b). Subsequently, I bred 10 chimeric animals (including the male mouse with 60% chimerism) with wild-type animals and genotyped the offspring to monitor transmission of the mutant allele. I tested two to three litters from each breading pair, but all the offspring (*F1* = 217) were wild type. To test whether KI mice embryos were dying during embryogenesis, I performed Sanger sequencing on DNA extracted from 18 embryos from two litters. None of them carried the introduced mutation. I carried out the same procedure on different tissues of chimeric males. Although I detected the introduced mutation in different tissues, I never detected it in testicular cells. This suggested that p.His373Arg homozygous ES cells do not contribute efficiently to the formation of chimeric mice and that they lack the viability to form germ cells. To overcome this problem, I opted to generate a heterozygous mES cell line and used this for blastocyst injection. To increase the likelihood of generating heterozygous mES cell lines, I slightly modified the experimental procedure. Briefly, I used two HRD templates: the one described above (c.1116C>T, c.1118A>G; HRD) and a second one, containing only a modified PAM sequence (c.1116C>T; HRD1). Again, I co-transfected mES cell line EDJ #22 with CRISPR Narf #3 and with both HRD templates (Figure 19a, right path) present in equal amounts during transfection. As described above, I selected cells via FACS, cultured RFP-positive colonies, and isolated and genotyped genomic DNA via Sanger sequencing. Of the 511 positive mES cell lines I obtained, 3 were homozygous, and 10 were compound heterozygous, but none were heterozygous clones. I repeated this experiment and co-transfected with a higher concentration of HRDs, which resulted in 276 RFP-positive colonies, including 4 homozygous, 6 compound heterozygous, and 3 heterozygous clones. three heterozygous knock-in mES (KI het) were injected into blastocysts, resulting in seven low-grade chimeras. Due to these chimeras' very low level of mosaicism and the resultant high likelihood of failed germline transmission of the mutation, I decided against subjecting them to further study. Therefore, I directly subjected the generated mES cell lines to subsequent *in vitro* analyses.

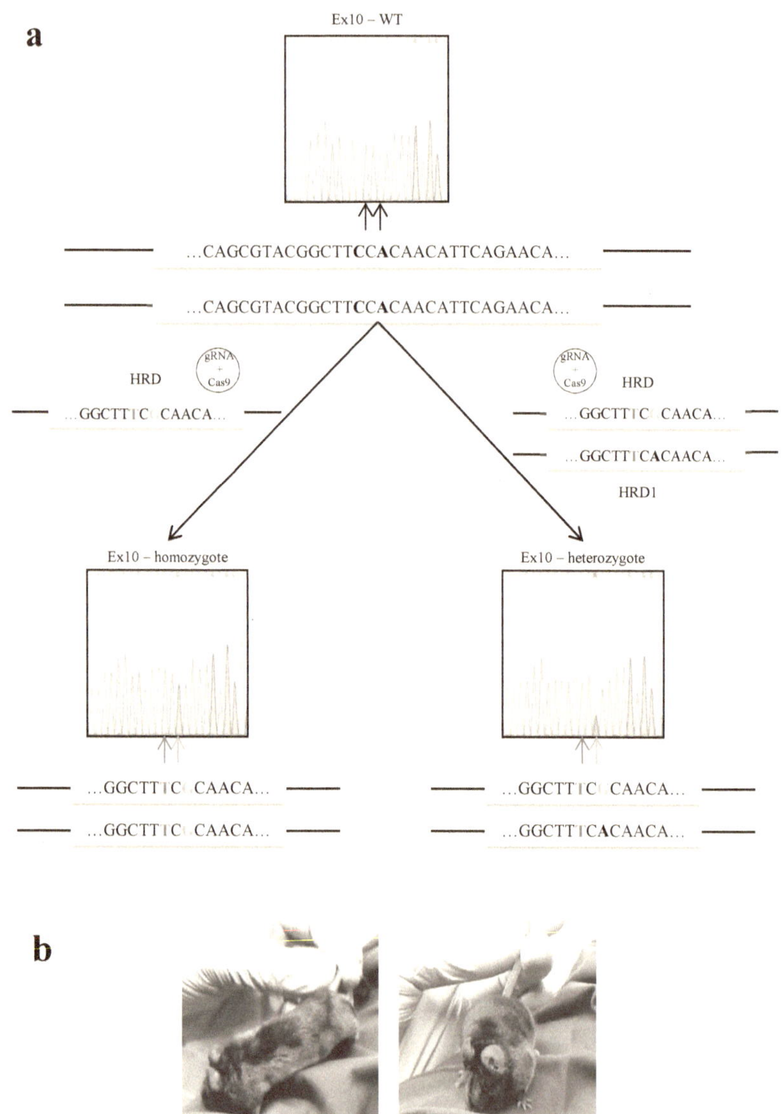

Figure 19: Generation of a p.H373R knock-in mouse model. (a) Schematic representation of strategy adopted for the generation of homozygous and heterozygous NARF[p.H367R]-KI mouse embryonic stem cells using CRISPR/Cas9 technology. After performing transfection with plasmid-encoding guided RNA, caspase 9 (Cas9), and a homologous recombinant DNA template, I obtained homozygous mES cells. The HRD contained a missense mutation corresponding to the mutation identified in the patient (c.1118A>G, green arrow) as well as a silent mutation in the PAM sequence (c.1116C>T, red arrow). To generate heterozygous mES cells, I used

additional HRD1. The HRD1 contained only a silent mutation in the PAM sequence. The chromatograms represent the results of Sanger sequencing of WT ES (upper) or KI mES cell lines, which I further used for blastocyst injection (lower). **(b)** Representative images of the 60% chimeric male obtained after the injection of homozygous Narf[p.H1373R] mES cells.

Due to the inability to carry out further tests *in vivo*, I characterised the *Narf* function and the mutational effect using the generated mES cell lines. I used one KI hom and one KI het mES line (further referred to as *Narf* KI hom mES cells and *Narf* KI het mES cells, respectively). EDJ #22 WT cells (further referred to as *Narf* WT mES cells) served as a control.

During culturing, I noticed differences between different clones in terms of the rate of cell proliferation. To analyse cell proliferation in detail, I performed a proliferation assay and measured changes in the number of cells within 48 hours after seeding the same number of cells. I then estimated the relative proliferation rates of the two tested cell lines and the control. The proliferation test indicated that *Narf* KI mES cells (both homozygous and heterozygous) proliferate significantly slower than *Narf* WT mES cells (Figure 20). This result suggested that the c.1118A>G mutation, whether in heterozygous or homozygous cells, impairs Narf function and engenders decreased proliferation.

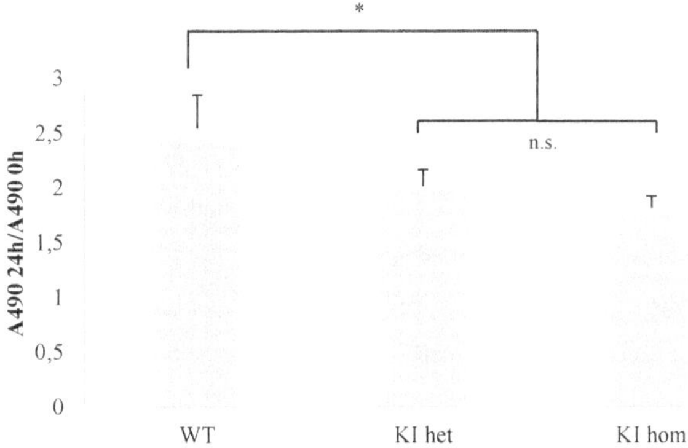

Figure 20: Proliferation assay in *Narf* KI mES cells. The cells reduce MTS tetrazolium into coloured formazan product which quantity can be measured. The quantity of formazan product reflecting the number of living cells was estimated by measuring of the absorbance at 490 nm with NanoDrop™ OneC Spectrophotometer. The KI mES cell lines (KI het and KI hom) proliferated significantly more slowly than the WT mES cells. The values and associated error bars represent the mean ± *SD* (*n*=3). The statistically significant values are indicated with asterisks (**$p<0.01$).

5.8   Genomic instability in *Narf* KI mES cells

Because genomic instability is one of the main and well-described hallmarks of ageing (Lopez-Otin et al., 2013) and exerts an influence on cell proliferation, I was highly interested in determining whether this mechanism could explain the phenotype identified in KI mES cells. For this purpose, I treated *Narf* WT, KI het, and KI hom mES cells with etoposide (Eto) or UV-C radiation (UV) in order to induce DNA damage, and I measured the efficiency of DNA repair mechanisms in these cells. Etoposide is known to introduce DSBs in the DNA, which can be repaired by NHEJ or HDR, while UV light exposure results in the generation of single nucleotide lesions that are mainly repaired by the nucleotide exchange repair (NER) mechanism. An impaired or inefficient NER mechanism can in turn lead to the generation of DSBs (Rastogi et al., 2010). Thus, I treated the cells with either etoposide or UV-C light and monitored the phosphorylation of histone H2AX (γH2AX), a well-established marker for the presence of DNA damage. Subsequently, I collected the cells at three different time points (1, 6, and 24 hours after the treatment/exposure) and submitted them for total protein extraction. Untreated cells served as the control in this experiment. Next, using WB analyses, I investigated the amount of γH2AX and total H2AX in the protein extracts. Etoposide treatment caused a dramatic increase in γH2AX levels within the first hour in all the different cell lines; the γH2AX levels then begin to slowly decrease over time (Figure 21a, upper panels). The relative values of the γH2AX amounts, estimated based on total H2AX expression, revealed no significant differences between cell lines in terms of the amounts of γH2AX (Figure 21b). UV radiation induces a gradual increase in γH2AX levels, which generally peak within six hours after exposure and then start to decline (Figure 21a, lower panels). The relative values of the γH2AX amounts, estimated as described above, indicated that the *Narf* KI hom mES cells exhibited higher γH2AX levels than the WT controls for a prolonged period; this suggested an impaired NER mechanism in this cell line (Figure 21c). The *Narf* KI het mES cells also presented elevated γH2AX levels 24 hours after treatment, although the difference was not statistically significant (Figure 21c).

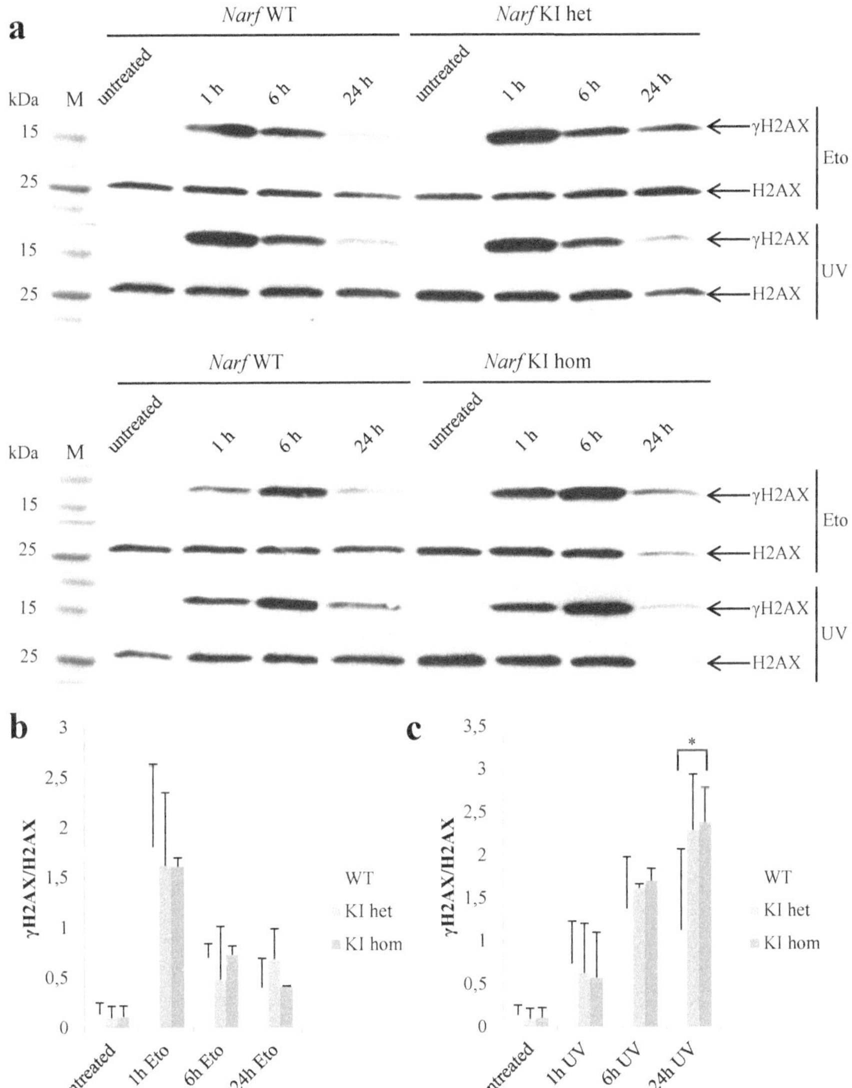

Figure 21: Genomic instability in *Narf* KI mES cell lines. (a) Representative immunoblot illustrating the expression of phosphorylated (γH2AX) and total H2AX protein in WT and KI (heterozygous and homozygous) mES cells 1, 6, and 24 hours after treatment with etoposide (Eto, upper panel) or UV-C (lower panel). **(b)** Using the ImageLab software, I quantified the γH2AX band intensities after etoposide treatment and normalised them to their respective total H2AX band intensities, which are represented as relative values in a bar graph. After etoposide treatment, γH2AX levels dramatically increased within 1 hour and began to decline over time (6 h, 24

h). There were no significant differences between cell lines in terms of the amounts of γH2AX observed (two-tailed *t*-test, *p*>0,05). **(c)** I quantified the γH2AX band intensities after UV-C exposure as described above. The UV-C irradiated cells exhibited a gradual increase in γH2AX levels within the first six hours after treatment; the γH2AX levels then began to decline. It was only in the *Narf* KI homozygous cells (KI hom) that I observed significant high levels of γH2AX in comparison to the WT cells (two-tailed *t*-test, **p*<0.05). The values and associated error bars represent the mean ± *SD* (*n*=2). M = Precision Plus Protein™ All Blue Pre-Stained Protein Standards.

In addition, I investigated whether knock-down of *NARF* influences genome stability in human fibroblasts. I generated human fibroblasts with knocked-down *NARF* gene through small interfering RNA (siRNA) (further referred to as *NARF* KD FB). I used the *NARF* KD FBs to the genomic stability analyses described above. The results demonstrated that fibroblasts in general are able to recover from DNA damage faster and more efficiently than ES cells: γH2AX appeared within 1 hour after etoposide treatment (Figure 22a, upper panels), and the UV light exposure led to a slow increase in the amount of γH2AX, which peaked 6 hours after treatment and was then poorly detectable after 24 hours (Figure 22a, lower panels). Evaluation of γH2AX band intensities revealed no differences between the *NARF* KD and control fibroblasts (Figure 22b).

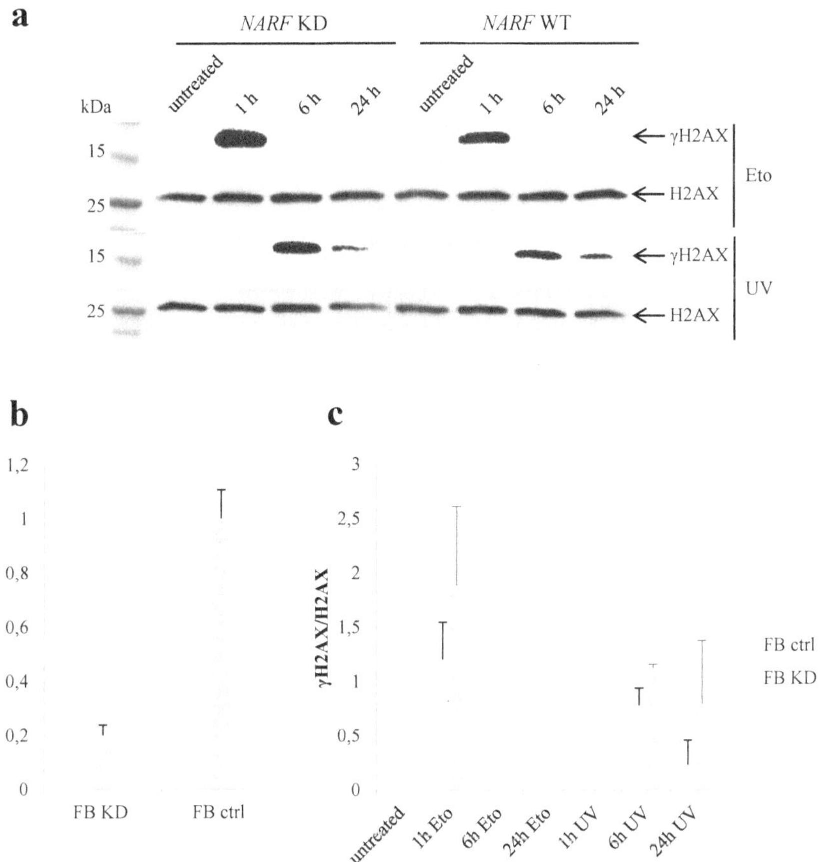

Figure 22: Genomic instability in *NARF* knocked-down fibroblasts. (a) WB results depicting the amount of γH2AX and total H2AX protein in *NARF* KD and control fibroblasts 1, 6, and 24 hours after treatment with etoposide (Eto) or exposure to UV-C radiation (UV). **(b)** qRT-PCR results demonstrating the efficiency of siRNA knock-down (KD) of *NARF* in fibroblasts at the time of experiment. siRNA transfection reduced the expression of *NARF* to ~20%. The values and associated error bars represent mean ± *SD* (*n*=3). **(c)** Using the ImageLab software, I quantified the γH2AX band intensities after Eto or UV treatment and normalised them to their respective total H2AX band intensities, which are represented as relative values in a bar graph. There were no significant differences between the *NARF* KD and control cells (two-tailed *t*-test, *p*>0.05). The values and associated error bars represent mean ± *SD* (*n*=2).

5.9   Oxidative stress response in NARF *in vitro* models

Since NARF homologues Nar1 and oxy-4 have been associated with oxidative stress in yeast and nematode, respectively (Fujii et al., 2009), I decided to investigate oxidative stress in NARF-compromised cells (mES cells and human fibroblasts). Therefore, I performed an oxidative stress response test using the commercially available CellROX™ Deep Red Flow Cytometry Assay Kit. Briefly, I induced the production of ROS in the cells, using tert-butyl hydroperoxide (TBHP). Next, I monitored the cellular response to this oxidative stress by staining the cells with CellROX® Deep Red reagent, which, as a reliable marker of ROS production in living cells, exhibits a strong fluorescent signal when oxidised. Simultaneously, I stained the cells with SYTOX® Blue Dead Cell to distinguish dead cells. I then measured and counted the fluorescent signals from both dyes by performing flow cytometry according to the manufacturer's protocol.

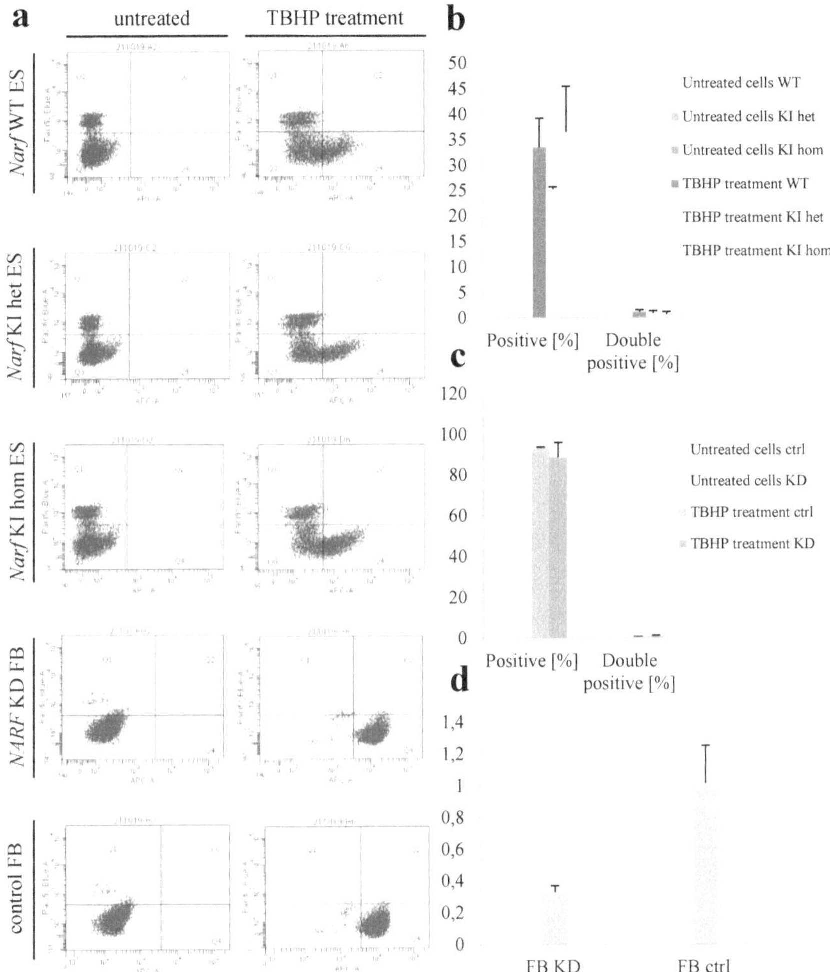

Figure 23: Oxidative stress assay in *Narf* KI mES cells and *NARF* KD fibroblasts. (a) Flow cytometry analysis of untreated and TBHP-treated WT cells, KI mES cells, WT fibroblasts, and *NARF* KD fibroblasts. CellROX® Deep Red (APC-A) positive cells represent ROS-positive cells, while SYTOX® Blue (Pacific Blue-A) positive cells represent apoptotic cells. **(b)** Evaluation of the flow cytometry results of the WT and KI mES cells. **(c)** Evaluation of the flow cytometry results of the WT and *NARF* KD fibroblasts. The untreated cells (both mES cells and fibroblasts) did not present any ROS-positive cells. Upon TBHP treatment, the ROS were activated, but there were no significant differences between treated cells and controls in terms of ROS generation or apoptosis (two-tailed *t*-test, $p > 0.05$). The values and associated error bars represent mean $\pm$ *SD* ($n=2$). **(d)** qRT-PCR results demonstrating the efficiency of siRNA knock-down (KD) of *NARF* in fibroblasts at the time of

experiment. siRNA transfection reduced the expression of *NARF* to ~30%. The values and associated error bars represent mean ± *SD* (*n*=3).

Without stimulation, none of the examined cell lines exhibited ROS accumulation (Figure 23a, untreated, Q4 squares). ES cells proved more sensitive to the staining procedures, as evinced in the higher number of apoptotic cells in the untreated cells than in the untreated fibroblasts, but no differences were observed between the *Narf* WT and KI mES cell lines or between the untreated and TBHP-treated cells (Figure 23a, Q1 squares). The TBHP treatment activated ROS production in all the cell lines in an equal manner, as compared to the WT controls (Figure 23a, TBHP treatment, Q4 squares). Induction of ROS production did not in itself lead to increased apoptosis, as double positive cells were barely detectable in all tested cell lines (Figure 23a, TBHP treatment, Q2 squares). When evaluating the flow cytometry results, I observed no significant differences between *Narf* KI mES cells and *Narf* WT mES cells or between *NARF* KD fibroblasts and control fibroblasts (Figure 23b, c, respectively) in terms of ROS accumulation (red positive cells) or apoptosis upon induction of oxidative stress (red/blue double positive cells). Using this assay, I was not able to demonstrate any impact of NARF on ROS accumulation or oxidative stress. Since the oxidative stress analysis was done at the end of my PhD-work, there was no time to perform additional and systematic analysis of oxidative stress testing.

6 Discussion

The main results obtained during my PhD project indicate that the cellular functions of NARF strongly depend on the proper localisation of protein within the nucleus. The mutation identified in patient prevents the nuclear translocation and causes cytoplasmic accumulation of protein. Functional analyses of protein interactions revealed novel insights into NARF's interaction partners. I confirmed interaction between NARF and lamin A and additionally determined two novel interactions – with CBX5 and NARF itself. Demonstrating the ability of NARF to form homodimers was a key step to discovering the dominant negative effect of the identified mutation. Interestingly, I was able to show that mutation does not disrupt NARF interactions; nevertheless, it impairs the nuclear import of the created dimers. Functional testing of Narf-mutant cells exhibited malfunction of some cellular properties which are also connected with the ageing-pathways. The mutation causes impaired cell proliferation abilities and dysfunction of DNA repair mechanisms leading to genomic instability. These two diminished cellular activities may be the main reason of the failure in *Narf* KI mouse line generation. At the same time, I was not able to determine functional similarities between NARF and its homologues. In addition, I could have not confirmed/excluded participation of NARF in the regulation of the oxidative stress mechanisms – a common function for all NARF homologues among different species.

6.1 Evolutionarily conserved function of NARF and its homologues

NARF is 31% identical and 41% similar to yeast Nar1. Due to this structural homology, it has been suggested that NARF performs a function similar to that of Nar1. A complementation assay performed in the study of Balk et al. (2004) was unable to prove this hypothesis (Balk et al., 2004). The authors tested both NARFL and NARF, but found that neither of them appears to have an evolutionarily conserved function. Since the authors were unable to present the expression and functionality of the NARF/NARFL proteins used in their complementation assay, I repeated this experiment. In my study, I was able to overexpress the human homologues in Nar1-depleted yeast, but they failed to rescue yeast growth. Systematic studies on yeast-human orthologous gene pairs have revealed that sequence similarity can be helpful in predicting functional rescue: sequence similarity is higher for complementing pairs (45%) than non-complementing pairs (29%) (Sun et al., 2016). At the same time, it has also been shown that the human non-orthologous protein may act as a functional substitute for the yeast protein as well (Hamza et al., 2015), suggesting that complementation cannot be completely predicted by sequence similarity (Hamza et al., 2015; Kachroo et al., 2015). Failure in the complementation assay could also be attributable to incorrect conformation of tagged human proteins, which in turn could cause the expression of non-functional proteins in yeast cells. As was demonstrated in this study, both the NARF and the NARFL proteins tagged with BFP2 exhibited untypical aggregation in the cytoplasm. Tagging of proteins with fluorescent epitopes can be helpful in determining the expression and proper localisation of the introduced protein; however, it can also lead to a decreased expression level and/or toxicity of the heterologous protein (Tugendreich et al., 2001). It is also known that complementation efficiency strongly depends on the cellular function of proteins. Systematic analyses indicate that a function-specific group of proteins can be almost completely replaceable (e.g. metabolic enzymes, which have over 90% replaceability), poorly replaceable (e.g. DNA replication and repair genes, which have 35% replaceability), or not replaceable at all (e.g. cell growth and dead genes, which have only 3% replaceability) (Kachroo et al., 2015), suggesting species-specific pathways that cannot be reproduced in different organisms. Considering that NARF contributes also to the maintenance of genomic stability and DNA repair, it falls into a group of proteins that is difficult to replace. Another species-specific feature that can prevent complementation is the protein-protein interactions network. This was illustrated in the example of human and yeast proteins involved in meiotic recombination, mitotic DNA repair, and telomere maintenance. The human *MRE11B* gene, which is a homologue of the yeast *MRE11* gene, was not able to rescue *mre11* mutants due to the lack of necessary interactions with yeast proteins (Chamankhah et al., 1998). Homologous proteins can contain conserved domains ensuring their conserved function among different organisms, but they can differ in terms of the remaining structure of the protein, and this part of the protein is species-specific. Such a situation has been described in proteins engaged in splicing and cell cycle control: Prp16 and Prp17, containing conserved WD repeats motifs in the C-terminal of proteins. The human *PRP16* and *PRP17* genes have both failed to exhibit the ability to rescue yeast *prp16* and *prp17* mutants, respectively. The generation of chimeric proteins composed of yeast N-terminal and human conserved C-terminal parts, however, has been found to result in functional replacement in both cases, indicating a species-specific role of the N-terminal part of yeast proteins (Ben Yehuda et al., 1998; Zhou and Reed, 1998). In the case of the examined Nar1, NARF, and NARFL homologues, all three share conserved C-terminal cysteine residues coordinating the H-cluster characteristic for iron-only hydrogenases (Balk et al., 2004). The generation of chimeric proteins containing the C-terminal H-cluster of NARF and NARFL fused with the yeast-specific N-terminus could potentially replace Nar1, culminating in improved functional complementation in yeast. Moreover, it has been proposed that selecting an adequate promoter for protein expression can be important in complementation assays. The commonly used constitutive or inducible promoters lead to artificial overexpression of the tested proteins; this, in turn, can disrupt the balance required for the proper protein function, thus leading to a lack of complementation (Lo Presti et al., 2009). In my study, all proteins used in the complementation assay were expressed from the inducible Met25 promoter. To rule out the possibility that suboptimal concentrations of NARF and NARFL block complementation of the Nar1 function in yeast, it may be advisable to try using native promoters. In general, there is no available set of common conditions that can ensure successful complementation, because each human-yeast gene pair is unique (Hamza et al., 2015). NARFL and Nar1 seem to be more closely related, as both are important players in the cytosolic iron-sulphur cluster assembly (CIA) machinery in mammals and yeast, respectively, but NARFL is still unable to substitute for the Nar1 function. The lower organisms, such as yeast or nematodes, possess only one homologue of this hydrogenase-like protein (Nar1 or oxy-4, respectively), whereas two (NARF and NARFL) are presented in higher eukaryotes, such as mammals. Many examples demonstrate that more complex organisms possess more than one homologue of essential genes. An excellent example is a group of genes involved in the NER mechanism, including the yeast *RAD6* gene and its human homologues, *HHR6A* and *HHR6B*, and the yeast *RAD23* gene and its human homologues, *HHR23A* and *HHR23B* (Koken et al., 1991; van der Spek et al., 1994). One can imagine that the *NAR1* gene might have split its function into two homologues during evolution. In this way, emerging proteins with structural homology can fulfil different functions. This would explain why neither NARF nor NARFL can replace the Nar1 function separately, raising the question of whether the two together would be able to replace the Nar1 function. Another interesting experiment would be to perform the complementation assay the other way around, using human and/or mouse cell lines lacking expression of the *NARF/Narf* and *NARFL/Narfl* genes to test the potential of the yeast *NAR1* or the nematode *OXY-4* gene to rescue the phenotype. Such an assay would allow for testing the ability of invertebrate proteins to complement mammalian proteins' functions. On the example of centromere proteins, it has been demonstrated that functional complementation can be examined in both ways because the yeast Cse4p homologue can substitute for a lack of the human CENP-A protein induced by RNAi in cells (Wieland et al., 2004).

6.2 Failing in the generation of specific NARF antibodies

There are two ways to immunise animals for antibody production: host animals can be injected either with the full-length protein or with short synthetic peptides (usually 10–20 amino acids in length; (Lee et al., 2016a). Both methods present advantages and disadvantages. Immunisation with the full-length protein allows for generating antibodies against multiple epitopes along the sequence. This in turn increases the chance that the generated antibodies will be able to detect at least one of these epitopes in native protein. At the same time, specificity to numerous epitopes can induce unspecific binding to other proteins possessing homologous sequences (pacificimmunology.com). The second strategy, immunisation with short peptides, ensures higher specificity of the produced antibodies (Trier et al., 2012), but in some cases, the target epitope is internally located in native protein, preventing the recognition and binding of the antibody (Lipman et al., 2005; Trier et al., 2012). In my approach, I tried to generate common antibodies against mouse and human NARF. For this purpose, I used two 16-amino-acid peptides that are complementary to both the mouse and the human protein. The generated antibodies passed all the internal quality controls; however, the specificity of the generated antibodies was tested on peptides used for immunisation but not on cell/tissue protein extracts. Since an antibody's specificity can be defined as its ability to recognise a particular epitope in the presence of other epitopes (Trier et al., 2012), the first specificity control was the WB analyses conducted in our lab. As illustrated in the results section, the antibodies were unable to recognise the endogenous NARF protein, instead displaying higher affinity to β-actin. In parallel, the generated antibodies recognised exogenous NARF protein artificially overexpressed in cells. The reason why they were unable to bind to the endogenous protein could relate to low quantity of endogenous NARF or the conformation of its native form, which may have precluded the antibodies from finding the specific epitope used for immunisation. The use of another peptide or the full-length protein for immunisation could perhaps enable successful antibody production. The 3D structure of the NARF protein would be helpful in selecting the optimal protein sequence for immunisation; otherwise, it has often been shown that exposed regions, such as N- or C-terminal parts, can constitute effective targets (Trier et al., 2012). Ultimately, it might be worth endeavouring to generate monoclonal antibodies (mAb), which provide homogeneity and high monospecificity (Lipman et al., 2005; Trier et al., 2012).

6.3 New insights into NARF interactions

To validate a previously described interaction between NARF and pre-lamin A (Barton and Worman, 1999) and to identify novel interaction partners of the NARF protein, I also conducted a Y2H experiment. In contrast to the previous report, the Y2H experiment conducted within this study failed to substantiate a NARF-pre-lamin A interaction. The main difference between my study and the study conducted by Barton and Worman (1999) is based on the direction of screening. Whereas Barton and Worman used lamin A as the bait, NARF served as the bait in my study. The other possible explanation for the disparity is that the two studies used different cDNA libraries. Barton and Worman (1999) used a cDNA library from HeLa cells (Barton and Worman, 1999), while I performed my screening on human ventricle and embryo heart cDNA libraries. I decided to use heart libraries because patient harbouring the NARF$^{p.H367R}$ mutation tend to display early onset of dilated cardiomyopathy. According to the Human Protein Atlas (proteinatlas.org), both NARF and lamin A are expressed in HeLa cells and heart muscle; however, it would be difficult to exclude the possibility that switching the assay direction or the library used in a Y2H assay can change the concentration and balance of the tested interaction partners, thereby influencing the interaction. The possibility that different outcomes originating in experiments conducted to search for interaction partners, especially Y2H assays, result from using different baits and libraries has already been published. Using a C-terminus of lamin A as the bait to screen a human skeletal muscle cDNA library identified COMMD1 as a novel binding partner of lamin A (Jiang et al., 2019), while screening the same cDNA library with full-length pre-lamin A as the bait allowed for identifying FAM96B as a lamin A interaction partner (Xiong et al., 2013). Using mouse lamin A has also resulted in identifying various interactions: a Y2H assay employing the C-terminus of mouse lamin A and the mouse 3T3-L1 adipocyte cDNA library uncovered interactions between lamin A and the Srebp1 and Srebp2 proteins (Lloyd et al., 2002). Interestingly, another group identified another lamin A binding partner, Sun1, using the same screening bait and library (Haque et al., 2006). These different interactions in different tissues may explain why mutations in one gene can inflict numerous forms of various disorders (Rankin and Ellard, 2006; Worman and Bonne, 2007), but they also hinder the identification of novel interaction partners. They may also explain why even more comprehensive studies adopting different systems have failed to identify NARF as an interaction partner of pre-lamin A/lamin A (Dittmer et al., 2014; Kubben et al., 2010). Nevertheless, I was able to validate a direct interaction between NARF and lamin A through CoIP and BiFC assays. I used two independent tests because the CoIP assay yielded unreproducible results. The technical difficulties associated with an IP experiment involving lamin A stem from its biochemical properties, which impede both the solubility of lamin A and the extraction of an intact protein complex. Conversely, stricter solubility conditions enable lamin A extraction, but disrupt interactions (Kubben et al., 2010).

The Y2H assay pointed to two novel interaction partners of NARF, one of which is CBX5. CBX5 (heterochromatin protein 1α – HP1α) is one of three non-histone chromatin-associated proteins belonging to the HP1 family (Lomberk et al., 2006). It controls heterochromatin organisation and gene silencing by binding to tri-methylated lysine 9 on histone H3 (H3K9me3) through the N-terminal chromo domain (CD) (Bannister et al., 2001; Lachner et al., 2001). Like lamin A, CBX5 has been shown to be involved in the accelerated ageing phenotype. Furthermore, it delayed DNA damage repair in *Zmpste24*-deficient mouse embryonic fibroblasts (MEFs) obtained from knock-out mice (Liu et al., 2014). Such findings render CBX5 an interesting partner for further investigation. In line with my study, more extensive proteomic analyses of HP1-binding proteins (HPBPs) revealed that CBX5 interacts with NARF (Nozawa et al., 2010). I confirmed this interaction *in vitro*, using two methods: a BiFC assay and a pull-down assay using recombinant CBX5 protein expressed in bacteria. I chose to use a pull-down assay to study this interaction due to similar difficulties with respect to the extraction of an intact NARF-CBX5 complex in CoIP experiments. Moreover, the pull-down assay revealed that the interaction between NARF and CBX5 is not disrupted by the p.H367R mutation in NARF, as I was able to demonstrate that mutant NARF also interacts with recombinant CBX5 *in vitro*. Using BiFC experiments, I attempted to examine the localisation of complexes of NARF$^{p.H367R}$ and both lamin A and CBX5; however, the results were inconclusive, revealing interacting proteins expressed in the nucleus and/or cytoplasm (data not shown). To overcome such ambiguous observations, it would be interesting to examine the localisation of these protein complexes over time because in all the BiFC assays, I fixed the cells and scrutinised them under the microscope at only one time point: 24 hours after transfection. A temporal analysis would offer insight into whether this dual localisation is attributable to dynamic processes of transfer of protein complexes or merely stems from the artificial situation created *in vitro*.

Most interestingly, the Y2H assay also uncovered a third possible interaction of NARF, which has not been previously described: a NARF-NARF interaction, which suggests the creation of a homodimer. This direct binding and formation of a homodimer was confirmed in the BiFC assay as well. Furthermore, the co-localisation of proteins in nuclear compartments indirectly corroborated all three of the discovered interactions. I found that NARF co-localises with lamin A within the nuclear envelope, but it is also expressed with CBX5 in the nucleoplasm.

6.4 Mislocalisation of NARF mutant protein and dominant negative effect

Unlike its cytoplasmic homologues, NARF has been described as a nuclear protein (Barton and Worman, 1999). In line with the result of a study conducted by Barton and Worman (1999), I identified overexpressed NARF in the nuclei of HeLa cells. Subsequently, I demonstrated that the patient-specific mutation causes mislocalisation of the NARF$^{p.H367R}$ protein presented exclusively in the cytoplasm. It was clearly established that mutations of the conserved histidine disturb the nuclear transport of the protein. Several pathways enable the nuclear import of proteins through nuclear pore complexes (NPCs): (1) canonical nuclear import mediated by importins α/β; (2) passive diffusion of small proteins; (3) travelling in a "piggyback" manner, using other interaction partners as transporters; (4) direct interaction with nucleoporins in the NPCs; (5) using the cytoskeleton to efficiently accumulate at the nuclear periphery, from where proteins can be easily carried by transporters; (6) using calmodulin/calreticulin as other transporting proteins in a calcium-dependent manner; (7) glycol-dependent transport via the leptin family of proteins (Bauer et al., 2015; Wagstaff and Jans, 2009). Active nuclear import requires the presence of nuclear localisation signal (NLS) in cargo proteins; NLS is recognised by importin α. Importin α subsequently interacts with importin β, acting as a linker between cargo proteins containing NLS and importin β, which ensures final transport via direct interactions with nucleoporins in the NPCs (Bauer et al., 2015; Christophe et al., 2000; Schlenstedt, 1996; Wagstaff and Jans, 2009). Since no classical NLS (cNLS) has been identified within the NARF sequence (as evaluated by cNLS Mapper, nls-mapper.iab.keio.ac.jp), it is important to consider other possibilities for the nuclear transport of NARF. Findings from the Y2H experiments indicated that NARF forms a homodimer. This insight might prove conducive to penetrating the process of the nuclear transfer of NARF, since I demonstrated that the p.H367R mutation exerts a dominant negative effect on the WT protein. Proteins very rarely operate as a single unit (Marianayagam et al., 2004; Matthews, 2012), and protein-protein interactions are instrumental in inducing most biological processes (Klemm et al., 1998). One of these widespread protein-protein interactions is dimerisation, which can be defined as an interaction between related subunits (Klemm et al., 1998). Dimerisation of proteins enhances the stability of proteins, increases the activity of enzymes by controlling active sites, and increases the specificity for nucleic acid binding sites in proteins engaged in DNA replication, DNA repair, and gene expression (Marianayagam et al., 2004; Mei et al., 2005). In addition, the formation of dimers requires less energy than long monomer synthesis and can assist in avoiding random associations (Mei et al., 2005). The generation of more complex structures also provides extended interaction surfaces, thus enabling simultaneous binding to other proteins and the formation of larger, better functioning complexes (Klemm et al., 1998; Marianayagam et al., 2004). Moreover, it has been revealed that the dimerisation of different types of proteins plays numerous roles in almost all signal transduction pathways (Klemm et al., 1998). The data collected here indicate that NARF performs its function in the nucleus most likely as a homodimer. Formation of this homodimer can also regulate its nuclear import or, at least, the generation of an unconventional 'structural' NLS. A similar situation has been described for the signal transducer and activator of transcription (STAT1). STAT1 nuclear localisation is ensured through tyrosine phosphorylation, followed by dimerisation. This dimerisation allows for generating an unconventional NLS that is subsequently recognised by importin-α5, resulting in nuclear import of the STAT1 dimer. Furthermore, it has been shown that the p.L497A mutation in STAT1 impairs the interaction of the STAT1 dimer with importin-α5 (and, in turn, impairs nuclear localisation), but does not impair phosphorylation, dimerisation, or DNA binding (McBride et al., 2002). These outcomes indicate that the NLS of the STAT1 protein becomes functional upon conformational changes that occur as a result of dimerisation (McBride et al., 2002). The functional heterogeneity of the importin α protein family and the capacity of its members to interact with multiple proteins via distinct domains (Miyamoto et al., 2016) support the possibility that dimerisation of the NARF protein also might generate a structural NLS that can be recognised by one of the family members. Conducting direct NARF interaction studies incorporating individual members of both the importin alpha and importin beta protein families could be helpful in determining whether these proteins are directly involved in the nuclear transport of NARF. Nevertheless, the possible presence of non-classical NLS within the NARF monomer cannot be completely excluded. The presence of unconventional NLS is reported frequently, on a case by case basis, and has been demonstrated in studies on particular nuclear proteins. Cloning approaches have also revealed some sequences that do not demonstrate specific characteristics but are partially similar to the sequences of known nuclear proteins, indicating that the pool of functional NLSs may be much larger than expected (Christophe et al., 2000). It would be worthwhile to explore the possibility of identifying new non-classical NLS within the NARF sequence. This could be accomplished by dividing the NARF protein into smaller parts containing specific domains or sequences that may function as NLS and fusing these sequences with a reporter protein, such as a fluorescent protein. Subsequently, the resultant fused fragments could be expressed in cells, and their ability to localise in nuclei could be evaluated under a microscope, especially for fragments including the highly important histidine at position 367. Such a process might reveal whether NLS-like sequences exist in the NARF protein. At the same time, NARF can also 'travel' to the nucleus in a 'piggyback' manner, using its interaction partners as transporters, since both lamin A and CBX5 possess cNLS sequences.

I demonstrated that mutations in NARF do not disrupt its interactions, or, at least, they do not disrupt its interaction with CBX5 *in vitro*. Unclear results regarding the localisation of mutant protein interactions, however, do not allow for evaluating whether mutations also change the location of NARF$^{p.H367R}$-partner complexes. As mentioned previously, repeating this experiment in a more precise, time-dependent manner could offer insight into this issue. Direct interaction studies could also be conducted to shed light on two other pathways of nuclear transport: interactions with NPCs and the cytoskeleton. The possibility that NARF is able to interact with nucleoporins, which build NPCs, and enters the nucleus via direct binding with nucleoporins should be explored. It is also worth examining whether NARF can use microtubular/actin filament movement to facilitate conventional nuclear import through accumulation at the nuclear periphery. Since there is no evidence for post-translational glycosylation of the NARF protein or its calcium-dependent function, it is unlikely that it can travel via calmodulin or leptin proteins. NARF is also too large to passively diffuse through the nuclear envelope. Evaluation of all the reviewed mechanisms of nuclear import can help to establish the particular pathway(s) that are used by NARF. It has been revealed that many proteins with critical nuclear roles have evolved the ability to use numerous mechanisms of nuclear transport that enable them to localise efficiently in the nucleus under different conditions, especially in situations where conventional nuclear import fails (Wagstaff and Jans, 2009). It is important, however, to heed the crucial role of dimerisation and the essential contribution of the histidine at position 367. It has been demonstrated that, due to pH-dependent changes in histidine chemistry (Li and Hong, 2011), histidine residues can play a critical role in protein dimerisation (Medina et al., 2019) and ligand binding/releasing (Rotzschke et al., 2002). Crystallographic examination of the three-dimensional (3D) structures of NARF dimers would be an excellent tool to identify conformational changes induced by histidine substitution. In addition to mislocalisation of the NARF$^{p.H367R}$ protein, it has been shown that mutations may cause reduced stability of the protein, resulting in accelerated degradation compared to WT NARF. Increased degradation of mislocalised proteins has been described for the transmembrane proteins that accumulate in the cytoplasm and cannot undergo proper folding (Hessa et al., 2011; Suzuki and Kawahara, 2016). The degradation of mislocalised NARF may suggest two possibilities: (1) nuclear localisation of the protein prevents its degradation, therefore mislocalised NARF is degraded in the cytoplasm, or (2) histidine substitution causes improper structure of the NARF dimer that not only disturbs nuclear transport but also becomes a signal for degradation. According to my results, however, decreased stability of NARF$^{p.H367R}$ can be observed in single-transfected HeLa cells. It could be interesting to explore the possibility that interaction with the WT protein in double-transfected cells can rescue mutant degradation.

6.5 Failing of Narf$^{p.H373R}$ mouse model generation

Another aim of my research project was to generate a Narf$^{p.H373R}$-KI mouse line as an *in vivo* model for functional study of *Narf*, a newly identified progeroid-associated gene. Gene editing by homologous recombination (HR) in mES cells has been a powerful tool to evaluate the function of a gene *in vivo*. The classical protocol for the generation of KI entails transfection of mES cells with a homologous recombination DNA template, which is used later for HR. The frequency of spontaneous recombination is extremely low and necessitates intense screening to identify positive clones. This renders the process extremely time-consuming (Sato et al., 2016). To facilitate and expedite the process, I decided to use the CRISPR/Cas9 system, which, combined with HRD templates, provides precise and specific HR at a specifically determined region in the genome (Jinek et al., 2012). I used the EDJ #22 mES cell line for mouse genome editing. As previously reported, this mES line contributes efficiently to chimera, leading to the generation of germline-transmitting male chimeras (Auerbach et al., 2000). The production of chimeras is a common intermediate step in the establishment of a mouse line, which additionally allows for following cell behaviour during both prenatal and postnatal development (Eakin and Hadjantonakis, 2006). To create a chimeric mouse, I employed one of the most common methods: injection of cells into the blastocoel cavity of a blastocyst stage embryo (Eakin and Hadjantonakis, 2006). This approach enabled to generate low-grade chimeras, which do not transmit KI alleles to offspring. Several key factors can affect the contribution of mES cells in the formation of germ cells, including the genetic background, the cell injection procedure itself, and chromosomal abnormalities and epigenetic alterations as a result of long-term culture and *in vitro* manipulation (Carstea et al., 2009). Furthermore, it is advisable to bear in mind that both, the generated *Narf* KI cell lines, which were injected into the blastocysts, carried a homozygous silent mutation in the PAM sequence. This variant was not associated with any changes in protein sequence; however, we are not able to predict its overall impact on gene function. Another cell line, carrying only this mutation, should be generated in the next trial of mouse model generation as a negative control. This will allow for controlling whether this change exerts an impact on genes behaviour and contribution to germ cell formation itself. In my first attempt at mouse line generation, I embraced the use of a homozygous *Narf* KI mES line, which failed in germ cell formation. This raised the assumption that a homozygous mutation might be deleterious for mES cells and prevent their proper contribution to proliferation and differentiation during development. To overcome these complications, I additionally used heterozygous *Narf* KI mES cells in the second trial. Unfortunately, the observed results were similar, as this cell line was able to create only poor, low-grade chimeras. This result might be due to the previously mentioned dominant negative effect of the introduced mutation and impaired proliferation of *Narf* KI mES cells.

Today the procedure of mouse model generation can be rendered easier and faster by eliminating particular steps required in conventional mES cell injection into blastocysts. Currently, instead of blastocysts, a two-cell stage zygote is used, and DNA modifications are introduced directly into the zygote through direct introduction of DNA, RNA, or ribonucleoproteins. The injection, in turn, has been replaced by other methods, such as microinjections, electroporation, viral transfection, or even direct delivery into male or female gonadal tissues (Gurumurthy and Lloyd, 2019; Sato et al., 2016). These advanced methods in genomic engineering have recently allowed for obtaining new efficient protocols for mouse gene editing, such as by CRISPR ribonucleoprotein (RNP) electroporation of zygotes (CRISPR-EZ) (Modzelewski et al., 2018). Moreover, it has been demonstrated that large gene cassettes can be introduced into the mouse genome using CRISPR RNP electroporation and adeno-associated virus (AAV) donor infection (CRISPR-READI) (Chen et al., 2019). This approach has been used to correct mouse hematopoietic stem and progenitor cells (mHSPCs) and restore B and T cell development *in vivo* (Tran et al., 2019). It would be worthwhile to consider using one of these new methods in future experiments to bypass the time-consuming and problematic nature of dealing with mES cells and chimeras. Nevertheless, the essential function of Narf might also be the reason that Narf-mutant mES cells do not contribute in the formation of cells from all the three primary germ layers.

6.6 Molecular function of NARF

6.6.1 Oxidative stress regulation as conserved function of hydrogenase-like proteins

As regulators of the oxidative stress response, all the identified NARF homologues share a common function. NARFL cooperates with hypoxia-inducible factor (HIF-1α) and serves as a regulator to maintain stable levels of HIF-1α. Knock-down of *NARFL* increases the expression of HIF-1α and its target genes under hypoxia and normoxia conditions (Huang et al., 2007). Recently, it has been reported that a homozygous mutation found in human *NARFL* caused pulmonary arteriovenous malformations (PAVMs) in two patients (Liu et al., 2017). A zebrafish model with *narfl* deletion revealed abnormal angiogenesis upon increased oxidative stress and upregulation of HIF-1α (Luo et al., 2019). Our patient with the *NARF* mutation, however, exhibited no malformations in her vascular system. Overexpression of NARFL in hyperoxia-resistant HeLa sublines also suggests that it plays a role in protecting both cytosolic and nuclear Fe-S proteins in hyperoxic environments (Corbin et al., 2015). Furthermore, a mutation found in *Caenorhabditis elegans* homologue *OXY-4* (*Y54H5A.4*), in the same region as the mutation found in our patient, has been shown to induce increased sensitivity to oxidative stress and decreased lifespan in worms (Fujii et al., 2009). Depletion of yeast Nar1 also leads to increased sensitivity to oxygen and lethality of yeast under hypoxic conditions (Fujii et al., 2009). The plant *NAR1* homologue has also been found to play an important role in the oxidative stress pathway in *Arabidopsis thaliana*; however, in this case, *nar1* mutants exhibited resistance to oxidative stress induced by paraquat (Nakamura et al., 2013). At the same time, an impaired response to oxidative stress has been reported in the fibroblasts of patients with atypical progeroid syndrome (APS)/atypical Werner syndrome (AWS) with a mutation in *LMNA* (Motegi et al., 2016). Nevertheless, in my study, I found that *Narf* mutations in mES cells exerted no influence on ROS production or on sensitivity to increased ROS levels in cells. I also found that *NARF* KD in human fibroblasts exerted no influence on the same variables. A general role of NARF in regulating the response to changing levels of oxygen, however, cannot be completely excluded on the basis of these outcomes. As has been demonstrated for nematode, yeast, and human cells, cultures in different oxygen concentrations can point to an impact on oxygen sensitivity. Therefore, it would be worthwhile to examine mES *Narf*-KIs and *NARF*-KD fibroblasts in both increased and decreased levels of oxygen and establish their behaviour under different oxygen conditions. Moreover, the presence of up- and down-regulated antioxidant enzymes and lower antioxidant defence has been revealed in AWS/WS fibroblasts (representing progeroid syndromes)

(Seco-Cervera et al., 2014). This downstream pathway of antioxidant defence against ROS accumulation should also be examined in in vitro models of *Narf* mutants. This can be accomplished by measuring the expression of antioxidant enzymes such as superoxide dismutase, catalase, thioredoxin, or glutaredoxin. It is important to evaluate the levels of these enzymes not only at the messenger RNA (mRNA) level but also at the protein level. It has been revealed, in an example of WRN cell lines from patients with Werner progeria syndrome, that gene expression and protein levels do not correlate completely and that posttranscriptional protein modification may also play an important role (Seco-Cervera et al., 2014).

6.6.2 Reduced cell proliferation capabilities

Both the *Narf* KI het and hom lines exhibited impaired proliferation caused by the introduced mutation. Similar outcomes have been observed for both Narfl, knock-down of which decreased the viability of mouse embryonic fibroblasts (MEFs) (Song and Lee, 2011), and Nar1, depletion of which led to yeast growth arrest (Balk et al., 2004). Impaired proliferation of cells carrying *Narf* mutations can explain some of the congenital and ageing features that presented in our patient. It has been reported that impaired proliferation e. g. can result in short stature. For example, it has been demonstrated that defects in chondrocyte proliferation, which is necessary for maintaining growth plate architecture and function, can result in a dwarfism phenotype (Koparir et al., 2015; Terpstra et al., 2003). Short stature is also a common characteristic in patients with WS and HGPS. The proliferative ageing that occurs in WS cells is connected with telomere shortening, whereas the proliferative ageing that occurs in HGPS occurs from defects in nuclear lamin, independent of telomerase activity (Zucchero and Ahmed, 2006). In addition to localisation in the nuclear lamina, lamin A locates in the nucleoplasm, in a complex with lamina-associated polypeptide (LAP2α). Loss of this complex increases cell proliferation, but in the presence of the progerin in HGPS cells, low LAP2α levels result in impaired proliferation. These data indicate that, depending on the level of lamin A in the nuclear interior, LAP2α can either promote or inhibit proliferation (Vidak et al., 2018). In cells lacking nuclear NARF, levels of nuceloplasmic lamin A might be impaired, resulting in a similar effect and a defect in proliferation. This may suggest a role of NARF in the localisation of its interacting partner lamin A within nuclear compartments. Moreover, it has been reported that the proliferation capacity of cells can be strongly dependent on their oxygen sensitivity and DNA damage repair abilities. It has been shown that mouse embryonic fibroblasts (MEFs) can proliferate more quickly under lower oxygen concentrations (3%) and accumulate more DNA damage in 20% oxygen. DNA damage in MEFs was also higher than DNA damage in human fibroblasts in 20% oxygen. This indicates that oxygen sensitivity can determine differences between mouse and human cell cultures and explain their proliferative differences *in vitro* (Parrinello et al., 2003). These data should be considered in future experiments involving mouse and human cells presenting *NARF* mutations. These findings point to the importance of distinguishing the role of NARF in oxidative stress response and cell proliferation or determining the overlap between these two pathways. Another of our patient's features that can be explained by impaired cell proliferation is microcephaly. This is a developmental brain anomaly that is rooted in defective proliferation of neuroprogenitors; it presents, for example, in patients with a mutation in *RTTN*. *RTTN* participates in cellular proliferation and neuronal migration and is involved in both isolated primary microcephaly and microcephalic primordial dwarfism (Shamseldin et al., 2015). Proliferation defects causing microcephaly have also been described in patients harbouring mutations in proteins involved in mitotic regulation and progression (Cavallin et al., 2017; Gilmore and Walsh, 2013; Sgourdou et al., 2017).

6.6.3 Impaired DNA damage repair and genomic instability

Genome stability is predominantly ensured by the proper functioning of the DNA damage repair mechanisms in cells. The cellular response to DNA damage is primarily based on initiation of the cascade of DNA damage response pathways (DDRs). DDRs are stimulated by lesion-specific sensor proteins that provide sufficient time for particular repair mechanisms to physically remove and exchange lesions. There are at least five major DNA repair pathways—base excision repair (BER), nucleotide excision repair, mismatch repair (MMR), homologous recombination (HR), and non-homologous end joining—which become activated at different cell cycle stages and respond to different types of damage (Chatterjee and Walker, 2017). To explore the possibility that NARF can regulate this process, researchers have tested genome stability in mES cells with *Narf* mutants (*Narf*-KI het and *Narf*-KI hom) by inducing DNA damage using etoposide and UV treatment to create DSBs and pyrimidine photoproducts (PPs), respectively (Montecucco and Biamonti, 2007; Sinha and Hader, 2002). Their results have indicated that the *Narf*-KI homozygous mutation affects the DNA repair mechanisms of UV-induced lesions, but not etoposide-induced DSBs. It is known that UV light irradiation engenders the creation of PPs, which are mostly repaired by NER, and NER malfunction can in turn lead to the generation of DSBs (Rastogi et al., 2010). A closer look at the NER mechanism reveals that adenosine triphosphate (ATP)-dependent DNA helicases xeroderma pigmentosum group D (XPD) or Fanconi anaemia complementation group J (FANCJ) are involved. XPD is a part of the 10-subunit complex transcription factor II human (TFIIH), which plays a role in both transcription initiation and in the NER pathway. FANCJ proteins are involved in the NER and Fanconi anaemia repair pathways, respectively (Rudolf et al., 2006). Dysfunction of these genes in humans causes xeroderma pigmentosum, Cockayne syndrome, trichothiodystrophy, or Fanconi anaemia (Lill, 2009; Lill and Muhlenhoff, 2008); some of these are described as progeroid syndromes. So far, the exact mechanism through which NARF regulates genome stability is unknown; however, both of its interaction partners, lamin A and CBX5, have been found to participate in DNA repair mechanisms. In the case of HGPS, defects in DNA repair pathways, telomere maintenance, epigenetic alterations, and oxidative stress all contribute to observed genomic instability (Gonzalo and Kreienkamp, 2015). Heterochromatin protein 1 (HP1) proteins regulate heterochromatin relaxation upon mobilisation via phosphorylation induced by DNA damage (Ayoub et al., 2008; Goodarzi et al., 2008; Ziv et al., 2006). In $Zmpste24^{-/-}$ cells lacking the zinc metalloproteinase which takes part in post-translational cleavage and maturation of lamin A, the phosphorylation of threonine at position 50 (pT50) and, in turn, the DNA damage response has been found to be significantly delayed due to the accumulation of pre-lamin A and the disorganisation of heterochromatin (Liu et al., 2014). Stabilisation of CBX5 through inhibition of its proteasomal degradation in a lamin-A-dependent manner has been described (Chaturvedi et al., 2012; Chaturvedi and Parnaik, 2010) There are studies indicating that pre-lamin A interacts with CBX5 and that the farnesylation of pre-lamin A decreases the capacity for this binding (Lattanzi et al., 2007), while NARF interacts with pre-lamin A in a farnesylated manner (Barton and Worman, 1999). Conversely, another study maintains that CBX5 is able to interact with pre-lamin A, lamin A, and lamin C (Liu et al., 2014), suggesting that binding sites for CBX5 and NARF are localised at different positions along the lamin A sequence. This raises a hypothesis that NARF could be a missing link between pre-lamin A and CBX5, thereby ensuring proper localisation inside the nucleus and binding of these proteins. In this case, mislocalisation of the NARF dimer prevents its functions in the nucleus.

All the experiments conducted in the in vitro model composed of *Narf*-KI het/hom mES cells and *NARF*-KD FB cell lines demonstrated that the introduced point mutation in both the heterozygous and homozygous states exerted a greater impact on cellular functions (proliferation, UV light sensitivity) than decreased expression of the *NARF* gene. It is important to mention, however, that reduced expression of the *NARF* mRNA was never complete and the lack/diminished expression of the protein was not confirmed due to the failure to generate anti-NARF antibodies and the absence of proper protein expression control.

6.7 Is NARF iron-sulphur protein?

As outlined in the introduction part, NARF and its homologues are similar to bacterial iron-only hydrogenases. Hydrogenases are enzymes mostly found in both anaerobic and aerobic prokaryotic organisms that can either produce molecular hydrogen (H_2) or use it as an energy source (Vignais et al., 2001; Wu and Mandrand, 1993). Occasionally, hydrogenases can be found in some anaerobic eukaryotic organisms, enabling their survival in an oxygenless environment (Degli Esposti et al., 2016; Horner et al., 2000; Horner et al., 2002). They catalyse a simple bidirectional conversion of hydrogen, protons, and electrons: $H_2 \leftrightarrow 2H^+ + 2e^-$. Based on the metal clusters at their catalytic sites, hydrogenases are classified into three major types: [NiFe]-, [FeFe]-, and metal-free-hydrogenases; however, most of the described hydrogenases belong to the first two families (Vignais et al., 2001; Winkler et al., 2013; Wu and Mandrand, 1993). Because the majority of eukaryotic organisms are aerobes and do not need to use hydrogenases for anaerobic processes, hydrogenase-like proteins have lost their enzymatic abilities and have been found to fulfil other functions in eukaryotic cells. All types of hydrogenases typically contain iron-sulphur (Fe-S) clusters, inorganic cofactors that bind to protein ligands. Fe-S clusters have the ability to accept or donate single electrons, to execute oxidation and reduction reactions, and to support electron transport. In most cases, Fe-S clusters are coordinated by conserved cysteine residues (rarely by histidine residues) to ensure effective conduction between the catalytic metal sites and external donors and/or acceptors of electrons (Peters et al., 2015). Iron-sulphur proteins are present in all living organisms; they were first reported and described in bacteria (Beinert, 2000; Beinert et al., 1997; Beinert and Thomson, 1983). On the path of evolution, iron-sulphur proteins were transferred by endosymbiosis to eukaryotic mitochondria, cytoplasms, and nuclei (Ciofi-Baffoni et al., 2018; Lill, 2009). Most Fe-S proteins are composed of either a rhomboid cluster of two iron and two sulphide ions [2Fe-2S] or a cuboidal cluster of four iron and four sulphide ions [4Fe-4S]. To preclude a toxic effect of free iron and sulphide, the assembly of Fe-S clusters is precisely controlled in cells. It seems that many steps of these processes are universal to all organisms, although Fe-S proteins' biosynthesis and maturation are much more complex in eukaryotes (Rouault, 2015). In eukaryotic cells, this process is split into mitochondrial iron-sulphur cluster (ISC) assembly machinery and cytosolic iron-sulphur cluster assembly (CIA) machinery. The ISC assembly machinery was inherited from endosymbiotic bacteria, and its components are homologues to the bacterial system (Ciofi-Baffoni et al., 2018; Lill, 2009). The processes of the mitochondrial ISC assembly system can be divided into three main functional steps: (1) *de novo* assembly of initial [2Fe-2S] cluster on ISCU2 scaffold protein (yeast Isu1), (2) transfer of [2Fe-2S] from ISCU2 to GLRX5 (Grx5), which functions as a [2Fe-2S] chaperon transferring the [2Fe-2S] to downstream acceptors, and (3) assembly of [4Fe-4S] cluster, followed by its insertion into mitochondrial protein. Within the first step, cysteine desulphurase complex NSF1-ISD11-ACP (Nsf1-Isd11-Acp1) removes sulphur from cysteine and transfers it to ISCU2 (Isu1). Next, the ISCU2 scaffold protein incorporates the iron from the mitochondrial pool to assemble the [2Fe-2S] cluster. In the next step, the *de novo* assembled [2Fe-2S] cluster is transferred from the ISCU2 scaffold protein to GLRX5 (Grx5). GLRX5 protects the [2Fe-2S] cluster and transfers it to final [2Fe-2S] acceptors and to proteins involved in the final step of the mitochondrial ISC, that is, maturation of the [4Fe-4S] target proteins. In the final step, ISCA1/ISCA2 (Isa1/Isa2) complex assembles the [4Fe-4S] cluster, which in turn is distributed to target proteins (Ciofi-Baffoni et al., 2018; Lill, 2009). ISC also has a fundamental function in the biosynthesis of cytosolic and nuclear Fe-S proteins. Disruption of the mitochondrial cysteine desulphurase complex Nfs1–Isd11 and the scaffold protein Isu1 in yeast has been found to be essential for extramitochondrial Fe/S protein biogenesis (Wiedemann et al., 2006). The components of ISC synthesise a sulphur-containing intermediate, called (Fe-S)int, which is transported to the cytoplasm and serves as a source for the cytosolic and nuclear Fe/S clusters (Pandey et al., 2019). In line with the hypothesised impaired function of Atm1, the mitochondrial-cytosol transporter of this (Fe-S)int component affects the biosynthesis of the cytosolic and nuclear Fe-S protein (Kispal et al., 1999). The (Fe-S)int component is further processed by the CIA machinery in the cytoplasm, fostering maturation of the cytosolic Fe-S protein and the transporter protein that supplies the Fe-S cluster to the cytosolic and nuclear target proteins. Exported (Fe-S)int is subsequently loaded on the NUBP1-NUBP2 scaffold protein complex (Cfd1-Nbp35). The scaffold protein complex uploads the Fe-S cluster into NARFL (Nar1), which together with CIAO1, CIA2B, MMS19, CIA2A, and CIAO1 (Cia1) forms the targeting complex (Ciofi-Baffoni et al., 2018; Lill, 2009). Depending on the composition of the targeting complex, the Fe-S cluster is transferred to either the cytosolic or the nuclear effector Fe-S protein. In contrast to mitochondrial Fe-S proteins, cytosolic and nuclear Fe-S proteins function as enzymes (aconitase (important component of the citric acid cycle) or biotin and lipoate synthases) (Booker et al., 2007; Meyer, 2008); regulate gene expression (cytosolic iron regulatory protein 1 (IRP1)) (Rouault, 2006); or control DNA repair mechanisms (ATP-dependent DNA helicases XPD or FANCJ) (Rudolf et al., 2006). Recent studies on Fe-S protein pathways have revealed that both Nar1 and NARFL play an important role in the CIA, regulating the maturation of cytosolic and nuclear but not mitochondrial Fe-S proteins (Balk et al., 2004; Song and Lee, 2008). It has been shown that Nar1 is crucial to yeast cells' function, as depletion of this protein causes growth arrest. Similarly, the knock-out of the *Narfl* gene is lethal in mice due to diminished activity of the cytosolic but not the mitochondrial Fe-S proteins (Song and Lee, 2011). In light of these results, I asked whether NARF can play a role in the maturation of the Fe-S proteins or is one of them. A comparative bioinformatics approach has also predicted that NARF is a Fe-S protein (Andreini et al., 2016), but this has not been proven experimentally. During my study, I investigated this issue through generation of the NARF protein in bacteria; I was able to purify recombinant NARF, but I failed to detect Fe-S components (data not shown). It could be that the conditions applied for recombinant NARF expression were not optimal for the production of Fe-S proteins. Briefly, I cultured BL21 *E.coli* bacteria in growth medium with Fe^{2+} and S^{2-}. After isopropyl β- d-1-thiogalactopyranoside (IPTG) induction, I extracted the recombinant protein under aerobic conditions. In a number of cases, supplementation with Fe^{2+} and S^{2-} was insufficient to induce expression of the Fe-S protein, and co-expression of the ISC components was essential for achieving high yields of active Fe-S holoprotein (Grawert et al., 2004; Kriek et al., 2003; Nakamura et al., 1999). At the same time, aerobic conditions would perhaps have converted the exposed Fe-S clusters to unstable forms that would have quickly decomposed (Imlay, 2006). The described modification should be implemented in the protocol so that the experiment could be repeated. If NARF is a true Fe-S protein, it could be the additional player in CIA. In contrast to NARFL, the knock-down of the NARF protein in HeLa and Hep3B cells exerted no impact on the maturation of the cytosolic Fe-S protein (Song and Lee, 2008). Nevertheless, it could still regulate the maturation of the nuclear Fe-S protein. So far, there is no information on how the Fe-S cluster is transported from the cytoplasm, through the nuclear membrane, into the nucleus. Since NARF is observed on the nuclear membrane, it would seem to be a strong candidate. The KI mES cells generated within my study will form a suitable model to address this question. As illustrated in the results section, the NARF[p.H367R] protein lost its nuclear localisation. If the NARF protein does in fact play a role in Fe-S transport, its mislocalisation should exert an impact on the function of the other Fe-S nuclear protein, e. g. DNA repair, maintenance of genomic stability, oxidative stress response or enzymatic activity.

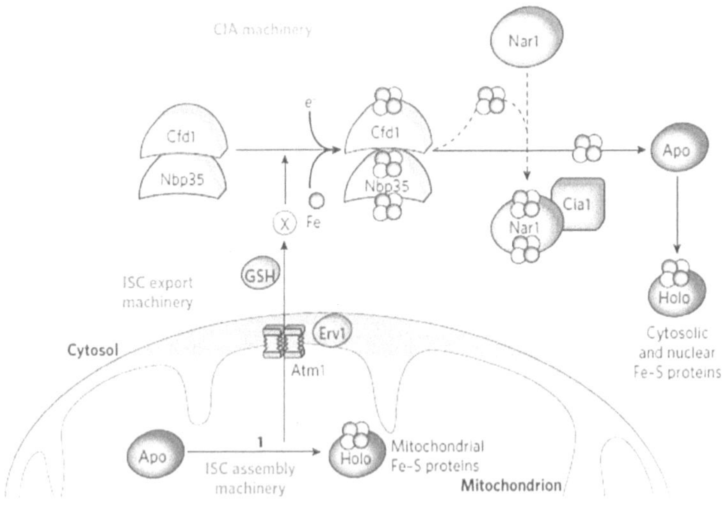

Figure 24: A model for biogenesis and maturation of cytosolic and nuclear Fe-S proteins in yeast. The Fe-S cluster intermediate is generated by mitochondrial iron-sulfur cluster (ISC) machinery and is exported from the mitochondrial matrix to the cytosol by Atm1. Cytosolic Fe-S protein assembly (CIA) machinery ensures transport and maturation of Fe-S proteins by assembly [Fe-S] cluster on scaffold protein complex Cfd1-Nbp35 and transfer complex to subsequent cluster carrier – Nar1. Nar1 interacting with Cia1 provides scaffold and target transfer of [Fe-S] cluster to proper apoproteins resulting in generation of mature cytosolic and nuclear holoproteins; 1 = shortcut of the three major steps of [Fe-S] cluster synthesis provided by mitochondrial ISC assembly machinery (adopted and modified from Lill, 2009).

6.8   Causative nature of p.His367Arg mutation in *NARF*

The evidence gathered during my PhD research strongly indicates the causative nature of the identified p.His367Arg mutation in *NARF*. This is evinced in many points, including the *de novo* occurrence of the patient's heterozygous mutation at a residue that is highly conserved among various species of yeast, invertebrates, and mammals. In addition, a very important point here is that the mutation resides at exactly the same amino acid residue as the nematode's homologue *OXY-4*, which causes increased sensitivity to environmental oxygen alterations and increased mortality among nematode mutants. The most interesting issue discovered during my studies is that the point substitution identified in *NARF* leads to complete mislocalisation of the protein. Besides, NARF is able to form dimers, and translocation of dimers to the nucleus is also impaired by the mutation. In addition, this mutation exerts a dominant negative effect on the wild-type protein, which may also explain the severe phenotype observed in a patient carrying the heterozygous form of the mutation. Furthermore, the mutation engenders dysfunction in cell proliferation abilities and impaired DNA damage repair mechanisms; these are both well-known hallmarks of ageing that also present in many progeroid syndromes. Finally, lamin A and CBX5, the interaction partners of NARF that were identified during the functional analyses conducted as part of this study, have also been associated with accelerated ageing disorders in humans and animal models. All the presented results provide genetic, functional, and morphological evidence for an important role of the identified mutation in the pathomechanisms of the progeroid syndrome afflicting our patient and indicate that the associated NARF variant may cause premature ageing diseases in humans.

7 List of figures

8   List of tables

List of tables

9 References

Addgene (2017). CRISPR 101: A Desktop Resource (2nd edition).

Alberts, B. (2002). Molecular biology of the cell, 4th edn (New York: Garland Science).

Andreini, C., Banci, L., and Rosato, A. (2016). Exploiting Bacterial Operons To Illuminate Human Iron-Sulfur Proteins. J Proteome Res *15*, 1308-1322.

Auerbach, W., Dunmore, J.H., Fairchild-Huntress, V., Fang, Q., Auerbach, A.B., Huszar, D., and Joyner, A.L. (2000). Establishment and chimera analysis of 129/SvEv- and C57BL/6-derived mouse embryonic stem cell lines. Biotechniques *29*, 1024-1028, 1030, 1032.

Ayoub, N., Jeyasekharan, A.D., Bernal, J.A., and Venkitaraman, A.R. (2008). HP1-beta mobilization promotes chromatin changes that initiate the DNA damage response. Nature *453*, 682-686.

Balk, J., Pierik, A.J., Aguilar Netz, D.J., Muhlenhoff, U., and Lill, R. (2004). The hydrogenase-like Nar1p is essential for maturation of cytosolic and nuclear iron–sulphur proteins. The EMBO Journal *23*, 2105–2115.

Bannister, A.J., Zegerman, P., Partridge, J.F., Miska, E.A., Thomas, J.O., Allshire, R.C., and Kouzarides, T. (2001). Selective recognition of methylated lysine 9 on histone H3 by the HP1 chromo domain. Nature *410*, 120-124.

Barrangou, R., Fremaux, C., Deveau, H., Richards, M., Boyaval, P., Moineau, S., Romero, D.A., and Horvath, P. (2007). CRISPR provides acquired resistance against viruses in prokaryotes. Science *315*, 1709-1712.

Barton, R.M., and Worman, H.J. (1999). Prenylated prelamin A interacts with Narf, a novel nuclear protein. J Biol Chem *274*, 30008-30018.

Bauer, N.C., Doetsch, P.W., and Corbett, A.H. (2015). Mechanisms Regulating Protein Localization. Traffic *16*, 1039-1061.

Behjati, S., and Tarpey, P.S. (2013). What is next generation sequencing? Arch Dis Child Educ Pract Ed *98*, 236-238.

Beinert, H. (2000). Iron-sulfur proteins: ancient structures, still full of surprises. J Biol Inorg Chem *5*, 2-15.

Beinert, H., Holm, R.H., and Munck, E. (1997). Iron-sulfur clusters: nature's modular, multipurpose structures. Science *277*, 653-659.

Beinert, H., and Thomson, A.J. (1983). Three-iron clusters in iron-sulfur proteins. Arch Biochem Biophys *222*, 333-361.

Belfort, M., and Roberts, R.J. (1997). Homing endonucleases: keeping the house in order. Nucleic Acids Res *25*, 3379-3388.

Ben Yehuda, S., Dix, I., Russell, C.S., Levy, S., Beggs, J.D., and Kupiec, M. (1998). Identification and functional analysis of hPRP17, the human homologue of the PRP17/CDC40 yeast gene involved in splicing and cell cycle control. RNA *4*, 1304-1312.

Bhaya, D., Davison, M., and Barrangou, R. (2011). CRISPR-Cas systems in bacteria and archaea: versatile small RNAs for adaptive defense and regulation. Annu Rev Genet *45*, 273-297.

Biolo, G., Heer, M., Narici, M., and Strollo, F. (2003). Microgravity as a model of ageing. Curr Opin Clin Nutr Metab Care *6*, 31-40.

References

Boch, J., and Bonas, U. (2010). Xanthomonas AvrBs3 family-type III effectors: discovery and function. Annu Rev Phytopathol *48*, 419-436.

Boch, J., Scholze, H., Schornack, S., Landgraf, A., Hahn, S., Kay, S., Lahaye, T., Nickstadt, A., and Bonas, U. (2009). Breaking the code of DNA binding specificity of TAL-type III effectors. Science *326*, 1509-1512.

Boeke, J.D., LaCroute, F., and Fink, G.R. (1984). A positive selection for mutants lacking orotidine-5'-phosphate decarboxylase activity in yeast: 5-fluoro-orotic acid resistance. Mol Gen Genet *197*, 345-346.

Bonjoch, L., Mur, P., Arnau-Collell, C., Vargas-Parra, G., Shamloo, B., Franch-Exposito, S., Pineda, M., Capella, G., Erman, B., and Castellvi-Bel, S. (2019). Approaches to functionally validate candidate genetic variants involved in colorectal cancer predisposition. Mol Aspects Med *69*, 27-40.

Booker, S.J., Cicchillo, R.M., and Grove, T.L. (2007). Self-sacrifice in radical S-adenosylmethionine proteins. Curr Opin Chem Biol *11*, 543-552.

Breslow, D.K., Cameron, D.M., Collins, S.R., Schuldiner, M., Stewart-Ornstein, J., Newman, H.W., Braun, S., Madhani, H.D., Krogan, N.J., and Weissman, J.S. (2008). A comprehensive strategy enabling high-resolution functional analysis of the yeast genome. Nat Methods *5*, 711-718.

Brouns, S.J., Jore, M.M., Lundgren, M., Westra, E.R., Slijkhuis, R.J., Snijders, A.P., Dickman, M.J., Makarova, K.S., Koonin, E.V., and van der Oost, J. (2008). Small CRISPR RNAs guide antiviral defense in prokaryotes. Science *321*, 960-964.

Burke, B., and Stewart, C.L. (2013). The nuclear lamins: flexibility in function. Nat Rev Mol Cell Biol *14*, 13-24.

Burtner, C.R., and Kennedy, B.K. (2010). Progeria syndromes and ageing: what is the connection? Nat Rev Mol Cell Biol *11*, 567-578.

Carrero, D., Soria-Valles, C., and Lopez-Otin, C. (2016). Hallmarks of progeroid syndromes: lessons from mice and reprogrammed cells. Dis Model Mech *9*, 719-735.

Carstea, A.C., Pirity, M.K., and Dinnyes, A. (2009). Germline competence of mouse ES and iPS cell lines: Chimera technologies and genetic background. World J Stem Cells *1*, 22-29.

Cavallin, M., Rujano, M.A., Bednarek, N., Medina-Cano, D., Bernabe Gelot, A., Drunat, S., Maillard, C., Garfa-Traore, M., Bole, C., Nitschke, P., *et al.* (2017). WDR81 mutations cause extreme microcephaly and impair mitotic progression in human fibroblasts and Drosophila neural stem cells. Brain *140*, 2597-2609.

Chamankhah, M., Wei, Y.F., and Xiao, W. (1998). Isolation of hMRE11B: failure to complement yeast mre11 defects due to species-specific protein interactions. Gene *225*, 107-116.

Chatterjee, N., and Walker, G.C. (2017). Mechanisms of DNA damage, repair, and mutagenesis. Environ Mol Mutagen *58*, 235-263.

Chaturvedi, P., Khanna, R., and Parnaik, V.K. (2012). Ubiquitin ligase RNF123 mediates degradation of heterochromatin protein 1alpha and beta in lamin A/C knock-down cells. PLoS One *7*, e47558.

Chaturvedi, P., and Parnaik, V.K. (2010). Lamin A rod domain mutants target heterochromatin protein 1alpha and beta for proteasomal degradation by activation of F-box protein, FBXW10. PLoS One *5*, e10620.

Chen, S., Sun, S., Moonen, D., Lee, C., Lee, A.Y., Schaffer, D.V., and He, L. (2019). CRISPR-READI: Efficient Generation of Knockin Mice by CRISPR RNP Electroporation and AAV Donor Infection. Cell Rep *27*, 3780-3789 e3784.

Chevalier, B.S., and Stoddard, B.L. (2001). Homing endonucleases: structural and functional insight into the catalysts of intron/intein mobility. Nucleic Acids Res *29*, 3757-3774.

Chojnowski, A., Ong, P.F., Wong, E.S., Lim, J.S., Mutalif, R.A., Navasankari, R., Dutta, B., Yang, H., Liow, Y.Y., Sze, S.K., *et al.* (2015). Progerin reduces LAP2alpha-telomere association in Hutchinson-Gilford progeria. Elife *4*.

Christophe, D., Christophe-Hobertus, C., and Pichon, B. (2000). Nuclear targeting of proteins: how many different signals? Cell Signal *12*, 337-341.

Ciofi-Baffoni, S., Nasta, V., and Banci, L. (2018). Protein networks in the maturation of human iron-sulfur proteins. Metallomics *10*, 49-72.

Clevers, H. (2016). Modeling Development and Disease with Organoids. Cell *165*, 1586-1597.

Corbin, M.V., Rockx, D.A., Oostra, A.B., Joenje, H., and Dorsman, J.C. (2015). The iron-sulfur cluster assembly network component NARFL is a key element in the cellular defense against oxidative stress. Free Radic Biol Med *89*, 863-872.

Dahl, K.N., Scaffidi, P., Islam, M.F., Yodh, A.G., Wilson, K.L., and Misteli, T. (2006). Distinct structural and mechanical properties of the nuclear lamina in Hutchinson-Gilford progeria syndrome. Proc Natl Acad Sci U S A *103*, 10271-10276.

Davies, B.S., Fong, L.G., Yang, S.H., Coffinier, C., and Young, S.G. (2009). The posttranslational processing of prelamin A and disease. Annu Rev Genomics Hum Genet *10*, 153-174.

De Sandre-Giovannoli, A., Bernard, R., Cau, P., Navarro, C., Amiel, J., Boccaccio, I., Lyonnet, S., Stewart, C.L., Munnich, A., Le Merrer, M., *et al.* (2003). Lamin a truncation in Hutchinson-Gilford progeria. Science *300*, 2055.

Dechat, T., Adam, S.A., Taimen, P., Shimi, T., and Goldman, R.D. (2010). Nuclear lamins. Cold Spring Harb Perspect Biol *2*, a000547.

Dechat, T., Pfleghaar, K., Sengupta, K., Shimi, T., Shumaker, D.K., Solimando, L., and Goldman, R.D. (2008). Nuclear lamins: major factors in the structural organization and function of the nucleus and chromatin. Genes Dev *22*, 832-853.

Decker, M.L., Chavez, E., Vulto, I., and Lansdorp, P.M. (2009). Telomere length in Hutchinson-Gilford progeria syndrome. Mech Ageing Dev *130*, 377-383.

Degli Esposti, M., Cortez, D., Lozano, L., Rasmussen, S., Nielsen, H.B., and Martinez Romero, E. (2016). Alpha proteobacterial ancestry of the [Fe-Fe]-hydrogenases in anaerobic eukaryotes. Biol Direct *11*, 34.

Deltcheva, E., Chylinski, K., Sharma, C.M., Gonzales, K., Chao, Y., Pirzada, Z.A., Eckert, M.R., Vogel, J., and Charpentier, E. (2011). CRISPR RNA maturation by trans-encoded small RNA and host factor RNase III. Nature *471*, 602-607.

Deng, D., Yan, C., Pan, X., Mahfouz, M., Wang, J., Zhu, J.K., Shi, Y., and Yan, N. (2012). Structural basis for sequence-specific recognition of DNA by TAL effectors. Science *335*, 720-723.

Dittmer, T.A., and Misteli, T. (2011). The lamin protein family. Genome Biol *12*, 222.

Dittmer, T.A., Sahni, N., Kubben, N., Hill, D.E., Vidal, M., Burgess, R.C., Roukos, V., and Misteli, T. (2014). Systematic identification of pathological lamin A interactors. Mol Biol Cell *25*, 1493-1510.

Duan, X., Gimble, F.S., and Quiocho, F.A. (1997). Crystal structure of PI-SceI, a homing endonuclease with protein splicing activity. Cell *89*, 555-564.

Duval, K., Grover, H., Han, L.H., Mou, Y., Pegoraro, A.F., Fredberg, J., and Chen, Z. (2017). Modeling Physiological Events in 2D vs. 3D Cell Culture. Physiology (Bethesda) *32*, 266-277.

Eakin, G.S., and Hadjantonakis, A.K. (2006). Production of chimeras by aggregation of embryonic stem cells with diploid or tetraploid mouse embryos. Nat Protoc *1*, 1145-1153.

Epinat, J.C., Arnould, S., Chames, P., Rochaix, P., Desfontaines, D., Puzin, C., Patin, A., Zanghellini, A., Paques, F., and Lacroix, E. (2003). A novel engineered meganuclease induces homologous recombination in yeast and mammalian cells. Nucleic Acids Res *31*, 2952-2962.

Eriksson, M., Brown, W.T., Gordon, L.B., Glynn, M.W., Singer, J., Scott, L., Erdos, M.R., Robbins, C.M., Moses, T.Y., Berglund, P., *et al.* (2003). Recurrent de novo point mutations in lamin A cause Hutchinson-Gilford progeria syndrome. Nature *423*, 293-298.

Fernandez, A., Josa, S., and Montoliu, L. (2017). A history of genome editing in mammals. Mamm Genome *28*, 237-246.

Flatt, T., and Partridge, L. (2018). Horizons in the evolution of aging. BMC Biol *16*, 93.

Flynn, P.J., and Reece, R.J. (1999). Activation of transcription by metabolic intermediates of the pyrimidine biosynthetic pathway. Mol Cell Biol *19*, 882-888.

Friedmann, T., and Roblin, R. (1972). Gene therapy for human genetic disease? Science *175*, 949-955.

Fu, Y., Sander, J.D., Reyon, D., Cascio, V.M., and Joung, J.K. (2014). Improving CRISPR-Cas nuclease specificity using truncated guide RNAs. Nat Biotechnol *32*, 279-284.

Fujii, M., Adachi, N., K., S., and D., A. (2009). [FeFe]-hydrogenase-like gene is involved in the regulation of sensitivity to oxygen in yeast and nematode. Genes to Cells, 457-468.

Garneau, J.E., Dupuis, M.E., Villion, M., Romero, D.A., Barrangou, R., Boyaval, P., Fremaux, C., Horvath, P., Magadan, A.H., and Moineau, S. (2010). The CRISPR/Cas bacterial immune system cleaves bacteriophage and plasmid DNA. Nature *468*, 67-71.

Garrido-Cardenas, J.A., Garcia-Maroto, F., Alvarez-Bermejo, J.A., and Manzano-Agugliaro, F. (2017). DNA Sequencing Sensors: An Overview. Sensors (Basel) *17*.

Gerace, L., Comeau, C., and Benson, M. (1984). Organization and modulation of nuclear lamina structure. J Cell Sci Suppl *1*, 137-160.

Gerace, L., and Huber, M.D. (2012). Nuclear lamina at the crossroads of the cytoplasm and nucleus. J Struct Biol *177*, 24-31.

Gilmore, E.C., and Walsh, C.A. (2013). Genetic causes of microcephaly and lessons for neuronal development. Wiley Interdiscip Rev Dev Biol *2*, 461-478.

Goldman, R.D., Gruenbaum, Y., Moir, R.D., Shumaker, D.K., and Spann, T.P. (2002). Nuclear lamins: building blocks of nuclear architecture. Genes Dev *16*, 533-547.

Goldman, R.D., Shumaker, D.K., Erdos, M.R., Eriksson, M., Goldman, A.E., Gordon, L.B., Gruenbaum, Y., Khuon, S., Mendez, M., Varga, R., *et al.* (2004). Accumulation of mutant lamin A causes progressive changes in nuclear architecture in Hutchinson-Gilford progeria syndrome. Proc Natl Acad Sci U S A *101*, 8963-8968.

Gonzalez, J.M., Pla, D., Perez-Sala, D., and Andres, V. (2011). A-type lamins and Hutchinson-Gilford progeria syndrome: pathogenesis and therapy. Front Biosci (Schol Ed) *3*, 1133-1146.

Gonzalo, S., and Kreienkamp, R. (2015). DNA repair defects and genome instability in Hutchinson-Gilford Progeria Syndrome. Curr Opin Cell Biol *34*, 75-83.

Goodarzi, A.A., Noon, A.T., Deckbar, D., Ziv, Y., Shiloh, Y., Lobrich, M., and Jeggo, P.A. (2008). ATM signaling facilitates repair of DNA double-strand breaks associated with heterochromatin. Mol Cell *31*, 167-177.

Gordon, L.B., Kleinman, M.E., Miller, D.T., Neuberg, D.S., Giobbie-Hurder, A., Gerhard-Herman, M., Smoot, L.B., Gordon, C.M., Cleveland, R., Snyder, B.D., *et al.* (2012). Clinical trial of a farnesyltransferase inhibitor in children with Hutchinson-Gilford progeria syndrome. Proc Natl Acad Sci U S A *109*, 16666-16671.

Gordon, L.B., Rothman, F.G., Lopez-Otin, C., and Misteli, T. (2014). Progeria: a paradigm for translational medicine. Cell *156*, 400-407.

Grawert, T., Kaiser, J., Zepeck, F., Laupitz, R., Hecht, S., Amslinger, S., Schramek, N., Schleicher, E., Weber, S., Haslbeck, M., *et al.* (2004). IspH protein of Escherichia coli: studies on iron-sulfur cluster implementation and catalysis. J Am Chem Soc *126*, 12847-12855.

Gurumurthy, C.B., and Lloyd, K.C.K. (2019). Generating mouse models for biomedical research: technological advances. Dis Model Mech *12*.

Hackstein, J.H. (2005). Eukaryotic Fe-hydrogenases -- old eukaryotic heritage or adaptive acquisitions? Biochem Soc Trans *33*, 47-50.

Hamza, A., Tammpere, E., Kofoed, M., Keong, C., Chiang, J., Giaever, G., Nislow, C., and Hieter, P. (2015). Complementation of Yeast Genes with Human Genes as an Experimental Platform for Functional Testing of Human Genetic Variants. Genetics *201*, 1263-1274.

Harman, D. (2003). The Free Radical Theory of Aging. Antioxidants & Redox Signaling *5*, 557-561.

Heath, P.J., Stephens, K.M., Monnat, R.J., Jr., and Stoddard, B.L. (1997). The structure of I-CreI, a group I intron-encoded homing endonuclease. Nat Struct Biol *4*, 468-476.

Hessa, T., Sharma, A., Mariappan, M., Eshleman, H.D., Gutierrez, E., and Hegde, R.S. (2011). Protein targeting and degradation are coupled for elimination of mislocalized proteins. Nature *475*, 394-397.

Horner, D.S., Foster, P.G., and Embley, T.M. (2000). Iron hydrogenases and the evolution of anaerobic eukaryotes. Mol Biol Evol *17*, 1695-1709.

Horner, D.S., Heil, B., Happe, T., and Embley, T.M. (2002). Iron hydrogenases--ancient enzymes in modern eukaryotes. Trends Biochem Sci *27*, 148-153.

Huang, J., Song, D., Flores, A., Zhao, Q., Mooney, S.M., Shaw, L.M., and Lee, F.S. (2007). IOP1, a novel hydrogenase-like protein that modulates hypoxia-inducible factor-1alpha activity. Biochem J *401*, 341-352.

Imlay, J.A. (2006). Iron-sulphur clusters and the problem with oxygen. Mol Microbiol *59*, 1073-1082.

Ishino, Y., Shinagawa, H., Makino, K., Amemura, M., and Nakata, A. (1987). Nucleotide sequence of the iap gene, responsible for alkaline phosphatase isozyme conversion in Escherichia coli, and identification of the gene product. J Bacteriol *169*, 5429-5433.

Jiang, Z., Chen, W., Zhou, J., Peng, Q., Zheng, H., Yuan, Y., Cui, H., Zhao, W., Sun, X., Zhou, Z., *et al.* (2019). Identification of COMMD1 as a novel lamin A binding partner. Mol Med Rep *20*, 1790-1796.

Jinek, M., Chylinski, K., Fonfara, I., Hauer, M., Doudna, J.A., and Charpentier, E. (2012). A programmable dual-RNA-guided DNA endonuclease in adaptive bacterial immunity. Science *337*, 816-821.

Jurica, M.S., Monnat, R.J., Jr., and Stoddard, B.L. (1998). DNA recognition and cleavage by the LAGLIDADG homing endonuclease I-CreI. Mol Cell *2*, 469-476.

Jurica, M.S., and Stoddard, B.L. (1999). Homing endonucleases: structure, function and evolution. Cell Mol Life Sci *55*, 1304-1326.

Kachroo, A.H., Laurent, J.M., Yellman, C.M., Meyer, A.G., Wilke, C.O., and Marcotte, E.M. (2015). Evolution. Systematic humanization of yeast genes reveals conserved functions and genetic modularity. Science *348*, 921-925.

Kerppola, T.K. (2006). Design and implementation of bimolecular fluorescence complementation (BiFC) assays for the visualization of protein interactions in living cells. Nat Protoc *1*, 1278-1286.

Kim, Y.G., Cha, J., and Chandrasegaran, S. (1996). Hybrid restriction enzymes: zinc finger fusions to Fok I cleavage domain. Proc Natl Acad Sci U S A *93*, 1156-1160.

Kim, Y.G., and Chandrasegaran, S. (1994). Chimeric restriction endonuclease. Proc Natl Acad Sci U S A *91*, 883-887.

Kispal, G., Csere, P., Prohl, C., and Lill, R. (1999). The mitochondrial proteins Atm1p and Nfs1p are essential for biogenesis of cytosolic Fe/S proteins. EMBO J *18*, 3981-3989.

Kleinstiver, B.P., Prew, M.S., Tsai, S.Q., Topkar, V.V., Nguyen, N.T., Zheng, Z., Gonzales, A.P., Li, Z., Peterson, R.T., Yeh, J.R., *et al.* (2015). Engineered CRISPR-Cas9 nucleases with altered PAM specificities. Nature *523*, 481-485.

Klemm, J.D., Schreiber, S.L., and Crabtree, G.R. (1998). Dimerization as a regulatory mechanism in signal transduction. Annu Rev Immunol *16*, 569-592.

Koken, M.H., Reynolds, P., Jaspers-Dekker, I., Prakash, L., Prakash, S., Bootsma, D., and Hoeijmakers, J.H. (1991). Structural and functional conservation of two human homologs of the yeast DNA repair gene RAD6. Proc Natl Acad Sci U S A *88*, 8865-8869.

Koks, S., Dogan, S., Tuna, B.G., Gonzalez-Navarro, H., Potter, P., and Vandenbroucke, R.E. (2016). Mouse models of ageing and their relevance to disease. Mech Ageing Dev *160*, 41-53.

Koparir, A., Karatas, O.F., Yuceturk, B., Yuksel, B., Bayrak, A.O., Gerdan, O.F., Sagiroglu, M.S., Gezdirici, A., Kirimtay, K., Selcuk, E., *et al.* (2015). Novel POC1A mutation in primordial dwarfism reveals new insights for centriole biogenesis. Hum Mol Genet *24*, 5378-5387.

Kowalski, J.C., and Derbyshire, V. (2002). Characterization of homing endonucleases. Methods *28*, 365-373.

Kriek, M., Peters, L., Takahashi, Y., and Roach, P.L. (2003). Effect of iron-sulfur cluster assembly proteins on the expression of Escherichia coli lipoic acid synthase. Protein Expr Purif *28*, 241-245.

Kubben, N., and Misteli, T. (2017). Shared molecular and cellular mechanisms of premature ageing and ageing-associated diseases. Nat Rev Mol Cell Biol *18*, 595-609.

Kubben, N., Voncken, J.W., Demmers, J., Calis, C., van Almen, G., Pinto, Y., and Misteli, T. (2010). Identification of differential protein interactors of lamin A and progerin. Nucleus *1*, 513-525.

Lachner, M., O'Carroll, D., Rea, S., Mechtler, K., and Jenuwein, T. (2001). Methylation of histone H3 lysine 9 creates a binding site for HP1 proteins. Nature *410*, 116-120.

Lattanzi, G., Columbaro, M., Mattioli, E., Cenni, V., Camozzi, D., Wehnert, M., Santi, S., Riccio, M., Del Coco, R., Maraldi, N.M., *et al.* (2007). Pre-Lamin A processing is linked to heterochromatin organization. J Cell Biochem *102*, 1149-1159.

Lee, B.S., Huang, J.S., Jayathilaka, L.P., Lee, J., and Gupta, S. (2016a). Antibody Production with Synthetic Peptides. Methods Mol Biol *1474*, 25-47.

Lee, J., Chung, J.H., Kim, H.M., Kim, D.W., and Kim, H. (2016b). Designed nucleases for targeted genome editing. Plant Biotechnol J *14*, 448-462.

Li, L., Wu, L.P., and Chandrasegaran, S. (1992). Functional domains in Fok I restriction endonuclease. Proc Natl Acad Sci U S A *89*, 4275-4279.

Li, S., and Hong, M. (2011). Protonation, tautomerization, and rotameric structure of histidine: a comprehensive study by magic-angle-spinning solid-state NMR. J Am Chem Soc *133*, 1534-1544.

Lill, R. (2009). Function and biogenesis of iron-sulphur proteins. Nature *460*, 831-838.

Lill, R., and Muhlenhoff, U. (2008). Maturation of iron-sulfur proteins in eukaryotes: mechanisms, connected processes, and diseases. Annu Rev Biochem *77*, 669-700.

Lipman, N.S., Jackson, L.R., Trudel, L.J., and Weis-Garcia, F. (2005). Monoclonal versus polyclonal antibodies: distinguishing characteristics, applications, and information resources. ILAR J *46*, 258-268.

Litke, R., Boulanger, E., and Fradin, C. (2018). [Caenorhabditis elegans as a model organism for aging: relevance, limitations and future]. Med Sci (Paris) *34*, 571-579.

Liu, B., Wang, J., Chan, K.M., Tjia, W.M., Deng, W., Guan, X., Huang, J.D., Li, K.M., Chau, P.Y., Chen, D.J., *et al.* (2005). Genomic instability in laminopathy-based premature aging. Nat Med *11*, 780-785.

Liu, H.Z., Du, C.X., Luo, J., Qiu, X.P., Li, Z.H., Lou, Q.Y., Yin, Z., and Zheng, F. (2017). A novel mutation in nuclear prelamin a recognition factor-like causes diffuse pulmonary arteriovenous malformations. Oncotarget *8*, 2708-2718.

Liu, J., Yin, X., Liu, B., Zheng, H., Zhou, G., Gong, L., Li, M., Li, X., Wang, Y., Hu, J., *et al.* (2014). HP1alpha mediates defective heterochromatin repair and accelerates senescence in Zmpste24-deficient cells. Cell Cycle *13*, 1237-1247.

Liu, L., Li, Y., Li, S., Hu, N., He, Y., Pong, R., Lin, D., Lu, L., and Law, M. (2012). Comparison of next-generation sequencing systems. J Biomed Biotechnol *2012*, 251364.

Livak, K.J., and Schmittgen, T.D. (2001). Analysis of relative gene expression data using real-time quantitative PCR and the 2(-Delta Delta C(T)) Method. Methods *25*, 402-408.

Lloyd, D.J., Trembath, R.C., and Shackleton, S. (2002). A novel interaction between lamin A and SREBP1: implications for partial lipodystrophy and other laminopathies. Hum Mol Genet *11*, 769-777.

Lo Presti, L., Cerutti, L., Monod, M., and Hauser, P.M. (2009). Choice of an adequate promoter for efficient complementation in Saccharomyces cerevisiae: a case study. Res Microbiol *160*, 380-388.

Lomberk, G., Wallrath, L., and Urrutia, R. (2006). The Heterochromatin Protein 1 family. Genome Biol *7*, 228.

Lopez-Otin, C., Blasco, M.A., Partridge, L., Serrano, M., and Kroemer, G. (2013). The hallmarks of aging. Cell *153*, 1194-1217.

Luo, J., Zhang, X., He, S., Lou, Q., Zhai, G., Shi, C., Yin, Z., and Zheng, F. (2019). Deletion of narfl leads to increased oxidative stress mediated abnormal angiogenesis and digestive organ defects in zebrafish. Redox Biol *28*, 101355.

Maeder, M.L., and Gersbach, C.A. (2016). Genome-editing Technologies for Gene and Cell Therapy. Mol Ther *24*, 430-446.

Marianayagam, N.J., Sunde, M., and Matthews, J.M. (2004). The power of two: protein dimerization in biology. Trends Biochem Sci *29*, 618-625.

Matthews, J.M. (2012). Protein dimerization and oligomerization in biology (New York Austin, Tex.: Springer Science+Business Media ; Landes Bioscience).

McBride, K.M., Banninger, G., McDonald, C., and Reich, N.C. (2002). Regulated nuclear import of the STAT1 transcription factor by direct binding of importin-alpha. EMBO J *21*, 1754-1763.

Medina, E., Villalobos, P., Conuecar, R., Ramirez-Sarmiento, C.A., and Babul, J. (2019). The protonation state of an evolutionarily conserved histidine modulates domainswapping stability of FoxP1. Sci Rep *9*, 5441.

Mei, G., Di Venere, A., Rosato, N., and Finazzi-Agro, A. (2005). The importance of being dimeric. FEBS J *272*, 16-27.

Meyer, J. (2008). Iron-sulfur protein folds, iron-sulfur chemistry, and evolution. J Biol Inorg Chem *13*, 157-170.

Miller, J., McLachlan, A.D., and Klug, A. (1985). Repetitive zinc-binding domains in the protein transcription factor IIIA from Xenopus oocytes. EMBO J *4*, 1609-1614.

Miller, J.C., Tan, S., Qiao, G., Barlow, K.A., Wang, J., Xia, D.F., Meng, X., Paschon, D.E., Leung, E., Hinkley, S.J., *et al.* (2011). A TALE nuclease architecture for efficient genome editing. Nat Biotechnol *29*, 143-148.

Miyamoto, Y., Yamada, K., and Yoneda, Y. (2016). Importin alpha: a key molecule in nuclear transport and non-transport functions. J Biochem *160*, 69-75.

Modzelewski, A.J., Chen, S., Willis, B.J., Lloyd, K.C.K., Wood, J.A., and He, L. (2018). Efficient mouse genome engineering by CRISPR-EZ technology. Nat Protoc *13*, 1253-1274.

References

Mojica, F.J., Diez-Villasenor, C., Garcia-Martinez, J., and Almendros, C. (2009). Short motif sequences determine the targets of the prokaryotic CRISPR defence system. Microbiology *155*, 733-740.

Montecucco, A., and Biamonti, G. (2007). Cellular response to etoposide treatment. Cancer Lett *252*, 9-18.

Moscou, M.J., and Bogdanove, A.J. (2009). A simple cipher governs DNA recognition by TAL effectors. Science *326*, 1501.

Motegi, S., Uchiyama, A., Yamada, K., Ogino, S., Yokoyama, Y., Perera, B., Takeuchi, Y., and Ishikawa, O. (2016). Increased susceptibility to oxidative stress- and ultraviolet A-induced apoptosis in fibroblasts in atypical progeroid syndrome/atypical Werner syndrome with LMNA mutation. Exp Dermatol *25 Suppl 3*, 20-27.

Mu, W., Lu, H.M., Chen, J., Li, S., and Elliott, A.M. (2016). Sanger Confirmation Is Required to Achieve Optimal Sensitivity and Specificity in Next-Generation Sequencing Panel Testing. J Mol Diagn *18*, 923-932.

Mumberg, D., Muller, R., and Funk, M. (1994). Regulatable promoters of Saccharomyces cerevisiae: comparison of transcriptional activity and their use for heterologous expression. Nucleic Acids Res *22*, 5767-5768.

Nakamura, M., Buzas, D.M., Kato, A., Fujita, M., Kurata, N., and Kinoshita, T. (2013). The role of Arabidopsis thaliana NAR1, a cytosolic iron-sulfur cluster assembly component, in gametophytic gene expression and oxidative stress responses in vegetative tissue. New Phytol *199*, 925-935.

Nakamura, M., Saeki, K., and Takahashi, Y. (1999). Hyperproduction of recombinant ferredoxins in escherichia coli by coexpression of the ORF1-ORF2-iscS-iscU-iscA-hscB-hs cA-fdx-ORF3 gene cluster. J Biochem *126*, 10-18.

Navarro, C.L., Cau, P., and Levy, N. (2006). Molecular bases of progeroid syndromes. Hum Mol Genet *15 Spec No 2*, R151-161.

Nozawa, R.S., Nagao, K., Masuda, H.T., Iwasaki, O., Hirota, T., Nozaki, N., Kimura, H., and Obuse, C. (2010). Human POGZ modulates dissociation of HP1alpha from mitotic chromosome arms through Aurora B activation. Nat Cell Biol *12*, 719-727.

Pandey, A.K., Pain, J., Dancis, A., and Pain, D. (2019). Mitochondria export iron-sulfur and sulfur intermediates to the cytoplasm for iron-sulfur cluster assembly and tRNA thiolation in yeast. J Biol Chem *294*, 9489-9502.

Pareek, C.S., Smoczynski, R., and Tretyn, A. (2011). Sequencing technologies and genome sequencing. J Appl Genet *52*, 413-435.

Parrinello, S., Samper, E., Krtolica, A., Goldstein, J., Melov, S., and Campisi, J. (2003). Oxygen sensitivity severely limits the replicative lifespan of murine fibroblasts. Nat Cell Biol *5*, 741-747.

Peters, J.W. (1999). Structure and mechanism of iron-only hydrogenases. Current Opinion in Structural Biology, 670-676.

Peters, J.W., Schut, G.J., Boyd, E.S., Mulder, D.W., Shepard, E.M., Broderick, J.B., King, P.W., and Adams, M.W. (2015). [FeFe]- and [NiFe]-hydrogenase diversity, mechanism, and maturation. Biochim Biophys Acta *1853*, 1350-1369.

Piper, M.D.W., and Partridge, L. (2018). Drosophila as a model for ageing. Biochim Biophys Acta Mol Basis Dis *1864*, 2707-2717.

Ponchel, F., Toomes, C., Bransfield, K., Leong, F.T., Douglas, S.H., Field, S.L., Bell, S.M., Combaret, V., Puisieux, A., Mighell, A.J., *et al.* (2003). Real-time PCR based on SYBR-Green I fluorescence: an alternative to the TaqMan assay for a relative quantification of gene rearrangements, gene amplifications and micro gene deletions. BMC Biotechnol *3*, 18.

Puca, A.A., Spinelli, C., Accardi, G., Villa, F., and Caruso, C. (2018). Centenarians as a model to discover genetic and epigenetic signatures of healthy ageing. Mech Ageing Dev *174*, 95-102.

Rankin, J., and Ellard, S. (2006). The laminopathies: a clinical review. Clin Genet *70*, 261-274.

Rastogi, R.P., Richa, Kumar, A., Tyagi, M.B., and Sinha, R.P. (2010). Molecular mechanisms of ultraviolet radiation-induced DNA damage and repair. J Nucleic Acids *2010*, 592980.

Rehm, H.L. (2013). Disease-targeted sequencing: a cornerstone in the clinic. Nat Rev Genet *14*, 295-300.

Rinaldi, F.C., Doyle, L.A., Stoddard, B.L., and Bogdanove, A.J. (2017). The effect of increasing numbers of repeats on TAL effector DNA binding specificity. Nucleic Acids Res *45*, 6960-6970.

Rotzschke, O., Lau, J.M., Hofstatter, M., Falk, K., and Strominger, J.L. (2002). A pH-sensitive histidine residue as control element for ligand release from HLA-DR molecules. Proc Natl Acad Sci U S A *99*, 16946-16950.

Rouault, T.A. (2006). The role of iron regulatory proteins in mammalian iron homeostasis and disease. Nat Chem Biol *2*, 406-414.

Rouault, T.A. (2015). Mammalian iron-sulphur proteins: novel insights into biogenesis and function. Nat Rev Mol Cell Biol *16*, 45-55.

Rudolf, J., Makrantoni, V., Ingledew, W.J., Stark, M.J., and White, M.F. (2006). The DNA repair helicases XPD and FancJ have essential iron-sulfur domains. Mol Cell *23*, 801-808.

Salomon, J.A., Wang, H., Freeman, M.K., Vos, T., Flaxman, A.D., Lopez, A.D., and Murray, C.J. (2012). Healthy life expectancy for 187 countries, 1990-2010: a systematic analysis for the Global Burden Disease Study 2010. Lancet *380*, 2144-2162.

Sanger, F., Nicklen, S., and Coulson, A.R. (1977). DNA sequencing with chain-terminating inhibitors. Proc Natl Acad Sci U S A *74*, 5463-5467.

Sapranauskas, R., Gasiunas, G., Fremaux, C., Barrangou, R., Horvath, P., and Siksnys, V. (2011). The Streptococcus thermophilus CRISPR/Cas system provides immunity in Escherichia coli. Nucleic Acids Res *39*, 9275-9282.

Sato, M., Ohtsuka, M., Watanabe, S., and Gurumurthy, C.B. (2016). Nucleic acids delivery methods for genome editing in zygotes and embryos: the old, the new, and the old-new. Biol Direct *11*, 16.

Saudi Mendeliome, G. (2015). Comprehensive gene panels provide advantages over clinical exome sequencing for Mendelian diseases. Genome Biol *16*, 134.

Schlenstedt, G. (1996). Protein import into the nucleus. FEBS Lett *389*, 75-79.

Schornack, S., Minsavage, G.V., Stall, R.E., Jones, J.B., and Lahaye, T. (2008). Characterization of AvrHah1, a novel AvrBs3-like effector from Xanthomonas gardneri with virulence and avirulence activity. New Phytol *179*, 546-556.

Schotik, M. (2017). Elucidating the function of novel genes and molecular mechanisms involved in accelerated ageing (Cologne: University of Cologne).

Seco-Cervera, M., Spis, M., Garcia-Gimenez, J.L., Ibanez-Cabellos, J.S., Velazquez-Ledesma, A., Esmoris, I., Banuls, S., Perez-Machado, G., and Pallardo, F.V. (2014). Oxidative stress and antioxidant response in fibroblasts from Werner and atypical Werner syndromes. Aging (Albany NY) *6*, 231-245.

Sgourdou, P., Mishra-Gorur, K., Saotome, I., Henagariu, O., Tuysuz, B., Campos, C., Ishigame, K., Giannikou, K., Quon, J.L., Sestan, N., *et al.* (2017). Disruptions in asymmetric centrosome inheritance and WDR62-Aurora kinase B interactions in primary microcephaly. Sci Rep *7*, 43708.

Shamseldin, H., Alazami, A.M., Manning, M., Hashem, A., Caluseiu, O., Tabarki, B., Esplin, E., Schelley, S., Innes, A.M., Parboosingh, J.S., *et al.* (2015). RTTN Mutations Cause Primary Microcephaly and Primordial Dwarfism in Humans. Am J Hum Genet *97*, 862-868.

Sieprath, T., Corne, T.D., Nooteboom, M., Grootaert, C., Rajkovic, A., Buysschaert, B., Robijns, J., Broers, J.L., Ramaekers, F.C., Koopman, W.J., *et al.* (2015). Sustained accumulation of prelamin A and depletion of lamin A/C both cause oxidative stress and mitochondrial dysfunction but induce different cell fates. Nucleus *6*, 236-246.

Sinha, J.K., Ghosh, S., and Raghunath, M. (2014). Progeria: a rare genetic premature ageing disorder. Indian J Med Res *139*, 667-674.

Sinha, R.P., and Hader, D.P. (2002). UV-induced DNA damage and repair: a review. Photochem Photobiol Sci *1*, 225-236.

Song, D., and Lee, F.S. (2008). A role for IOP1 in mammalian cytosolic iron-sulfur protein biogenesis. J Biol Chem *283*, 9231-9238.

Song, D., and Lee, F.S. (2011). Mouse knock-out of IOP1 protein reveals its essential role in mammalian cytosolic iron-sulfur protein biogenesis. J Biol Chem *286*, 15797-15805.

Song, D., Tu, Z., and Lee, F.S. (2009). Human ISCA1 interacts with IOP1/NARFL and functions in both cytosolic and mitochondrial iron-sulfur protein biogenesis. J Biol Chem *284*, 35297-35307.

Sun, S., Yang, F., Tan, G., Costanzo, M., Oughtred, R., Hirschman, J., Theesfeld, C.L., Bansal, P., Sahni, N., Yi, S., *et al.* (2016). An extended set of yeast-based functional assays accurately identifies human disease mutations. Genome Res *26*, 670-680.

Suzuki, R., and Kawahara, H. (2016). UBQLN4 recognizes mislocalized transmembrane domain proteins and targets these to proteasomal degradation. EMBO Rep *17*, 842-857.

Takata, M., Sasaki, M.S., Sonoda, E., Morrison, C., Hashimoto, M., Utsumi, H., Yamaguchi-Iwai, Y., Shinohara, A., and Takeda, S. (1998). Homologous recombination and non-homologous end-joining pathways of DNA double-strand break repair have overlapping roles in the maintenance of chromosomal integrity in vertebrate cells. EMBO J *17*, 5497-5508.

Tang, H., and Thomas, P.D. (2016). Tools for Predicting the Functional Impact of Nonsynonymous Genetic Variation. Genetics *203*, 635-647.

Terns, M.P., and Terns, R.M. (2011). CRISPR-based adaptive immune systems. Curr Opin Microbiol *14*, 321-327.

References

Terpstra, L., Prud'homme, J., Arabian, A., Takeda, S., Karsenty, G., Dedhar, S., and St-Arnaud, R. (2003). Reduced chondrocyte proliferation and chondrodysplasia in mice lacking the integrin-linked kinase in chondrocytes. J Cell Biol *162*, 139-148.

Thierry, A., and Dujon, B. (1992). Nested chromosomal fragmentation in yeast using the meganuclease I-Sce I: a new method for physical mapping of eukaryotic genomes. Nucleic Acids Res *20*, 5625-5631.

Tran, N.T., Sommermann, T., Graf, R., Trombke, J., Pempe, J., Petsch, K., Kuhn, R., Rajewsky, K., and Chu, V.T. (2019). Efficient CRISPR/Cas9-Mediated Gene Knockin in Mouse Hematopoietic Stem and Progenitor Cells. Cell Rep *28*, 3510-3522 e3515.

Trier, N.H., Hansen, P.R., and Houen, G. (2012). Production and characterization of peptide antibodies. Methods *56*, 136-144.

Tugendreich, S., Perkins, E., Couto, J., Barthmaier, P., Sun, D., Tang, S., Tulac, S., Nguyen, A., Yeh, E., Mays, A., *et al.* (2001). A streamlined process to phenotypically profile heterologous cDNAs in parallel using yeast cell-based assays. Genome Res *11*, 1899-1912.

Tupler, R., Perini, G., and Green, M.R. (2001). Expressing the human genome. Nature *409*, 832-833.

United Nations, D.o.E.a.S.A., Population Division (2017). World Population Ageing 2017 - Highlights (ST/ESA/SER.A/397) (New York: United Nations).

van der Spek, P.J., Smit, E.M., Beverloo, H.B., Sugasawa, K., Masutani, C., Hanaoka, F., Hoeijmakers, J.H., and Hagemeijer, A. (1994). Chromosomal localization of three repair genes: the xeroderma pigmentosum group C gene and two human homologs of yeast RAD23. Genomics *23*, 651-658.

Varela, I., Pereira, S., Ugalde, A.P., Navarro, C.L., Suarez, M.F., Cau, P., Cadinanos, J., Osorio, F.G., Foray, N., Cobo, J., *et al.* (2008). Combined treatment with statins and aminobisphosphonates extends longevlty in a mouse model of human premature aging. Nat Med *14*, 767-772.

Vidak, S., and Foisner, R. (2016). Molecular insights into the premature aging disease progeria. Histochem Cell Biol *145*, 401-417.

Vidak, S., Georgiou, K., Fichtinger, P., Naetar, N., Dechat, T., and Foisner, R. (2018). Nucleoplasmic lamins define growth-regulating functions of lamina-associated polypeptide 2alpha in progeria cells. J Cell Sci *131*.

Vignais, P.M., Billoud, B., and Meyer, J. (2001). Classification and phylogeny of hydrogenases. FEMS Microbiol Rev *25*, 455-501.

Wagstaff, K.M., and Jans, D.A. (2009). Importins and beyond: non-conventional nuclear transport mechanisms. Traffic *10*, 1188-1198.

Weaver, D.T. (1995). What to do at an end: DNA double-strand-break repair. Trends Genet *11*, 388-392.

Wiedemann, N., Urzica, E., Guiard, B., Muller, H., Lohaus, C., Meyer, H.E., Ryan, M.T., Meisinger, C., Muhlenhoff, U., Lill, R., *et al.* (2006). Essential role of Isd11 in mitochondrial iron-sulfur cluster synthesis on Isu scaffold proteins. EMBO J *25*, 184-195.

Wiedenheft, B., Sternberg, S.H., and Doudna, J.A. (2012). RNA-guided genetic silencing systems in bacteria and archaea. Nature *482*, 331-338.

References

Wieland, G., Orthaus, S., Ohndorf, S., Diekmann, S., and Hemmerich, P. (2004). Functional complementation of human centromere protein A (CENP-A) by Cse4p from Saccharomyces cerevisiae. Mol Cell Biol *24*, 6620-6630.

Winkler, M., Esselborn, J., and Happe, T. (2013). Molecular basis of [FeFe]-hydrogenase function: an insight into the complex interplay between protein and catalytic cofactor. Biochim Biophys Acta *1827*, 974-985.

World Health Organization (2017). Global strategy and action plan on ageing and health. (Geneva: World Health Organization).

Worman, H.J., and Bonne, G. (2007). "Laminopathies": a wide spectrum of human diseases. Exp Cell Res *313*, 2121-2133.

Wu, L.F., and Mandrand, M.A. (1993). Microbial hydrogenases: primary structure, classification, signatures and phylogeny. FEMS Microbiol Rev *10*, 243-269.

Zhou, Z., and Reed, R. (1998). Human homologs of yeast prp16 and prp17 reveal conservation of the mechanism for catalytic step II of pre-mRNA splicing. EMBO J *17*, 2095-2106.

Ziv, Y., Bielopolski, D., Galanty, Y., Lukas, C., Taya, Y., Schultz, D.C., Lukas, J., Bekker-Jensen, S., Bartek, J., and Shiloh, Y. (2006). Chromatin relaxation in response to DNA double-strand breaks is modulated by a novel ATM- and KAP-1 dependent pathway. Nat Cell Biol *8*, 870-876.

Zucchero, T., and Ahmed, S. (2006). Genetics of proliferative aging. Exp Gerontol *41*, 992-1000.